Cracking the Case: Understanding the Threat of Alkali-Silica Reaction in Concrete

Purky

Table of Contents

1.1 Alkali silica reaction (ASR) and aging structures

Alkali-silica reaction (ASR) is one of two types of alkali-aggregate reaction (AAR) in concrete. AAR was first discovered by Stanton in California in 1940 [1], after which it was reported in over 50 countries around the world [2]. Shortly after, the Hurdman bridge in Ottawa (Canada) was diagnosed with AAR and later demolished in 1987 [2]. The reaction is often characterized by a random pattern of cracks (often referred to as 'map-cracking') visible on the surface of affected concrete as shown in Figure 1.1. Portland Cement (PC) contains a given amount of alkalis which differs as per the PC type, which is present in the concrete pore solution in the form of hydroxides (i.e., Na^+, K^+-OH^-). In certain concrete mixtures, aggregates (fine or coarse) may contain unstable mineral phases, which may be attacked by alkali hydroxides. The above reaction is known as AAR in general, and ASR in when this reaction takes place in the presence of siliceous mineral phases [3]. ASR produces a secondary viscous product known as 'ASR gel' or 'silica gel', which swells upon moisture uptake, resulting in induced expansion and the generation of microcracks due to localized tensile stresses. As the reaction proceeds into latter stages, cracks grow in size (i.e., length and width), linking together to form a crack network which has significant implications on the material properties [4,5] of affected concrete (Figure 1.2). Thus, over the long term, the performance of affected concrete structures may be considerably diminished in relation to sound concrete (i.e., important deflections, reduced capacity for service loads, etc.). The above conclusion has raised many concerns among researchers and engineers regarding the safety and serviceability of aging infrastructures.

Figure 1.1: Map-cracking on the surface of AAR-damaged concrete [6].

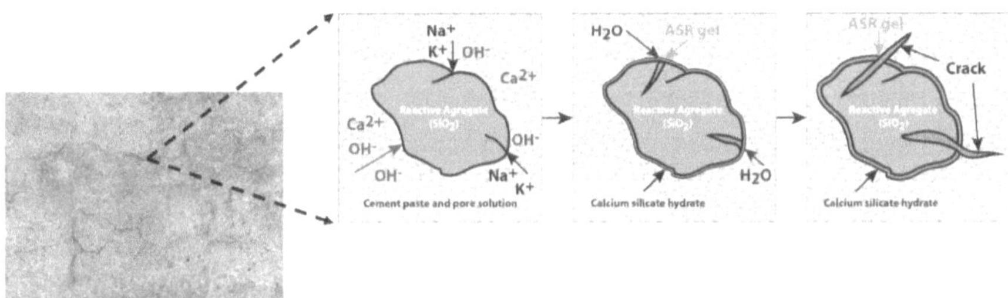

Figure 1.2: Mechanism of ASR in concrete [7].

Since the discovery of AAR, decades of research on reaction mechanisms, contributing factors and other aspects have led to the development of methods for the evaluation of alkali-reactivity [8–12], prevention of the reaction [7,13], and diagnosis and prognosis procedures and testing protocols [14–19]. The latter is quite significant in relation to the appraisal of existing infrastructures suffering from ASR. Although quite rare, past structures have been demolished as a consequence of the delayed identification and inefficient management of concrete distress. Therefore, an efficient diagnosis and prognosis plan is not only economically efficient, but more importantly it is the safest precaution in terms of ensuring public safety and prolonged serviceability for aging structures. Furthermore, the above may present the opportunity for the effective management of damage in the long term. However, for such procedures to be adopted more commonly in practice, it is important to gauge their reliability in a research setting first.

1.2   Research objectives

The aim of the following research is the appraisal (i.e., diagnosis) and prediction (i.e., prognosis) of ASR-induced expansion and deterioration of a concrete structure located at the University of Ottawa campus (i.e., SITE building) that demonstrates ASR deterioration after nearly 20 years of service. The first step of the diagnosis phase of the project involves conducting a visual inspection for the identification of potential ASR in various elements of the affected structure. Based on the above, the most damaged elements were selected for core extraction, in order to conduct a laboratory investigation program comprised of microscopic and mechanical testing. Afterwards, results were used to validate the type and extent of distress present in the structure. Then, a novel model was developed and validated using existing data and implemented to appraise the potential of future deterioration of the affected structure. The model was designed considering both microscopic (i.e., aggregate lithotype, temperature, relative humidity, alkali content) and macroscopic (i.e., geometry, reinforcement confinement, stress confinement) aspects of ASR, enabling its use for the prediction of damage on the structural level.

1.3   book organization

The present book is composed of five chapters. The first chapter is the introduction, which presents a brief background on ASR and its impact on affected infrastructure. The second chapter displays a thorough background and literature review on ASR in aging structures along with an evaluation of testing procedures and protocols. In the above, findings from previous works demonstrating the influence of several factors (i.e., both microscopic and macroscopic) are discussed to establish a background for the following chapters. Furthermore, non-destructive and destructive test methods are presented as potential diagnostic procedures to be used in the case study. The third chapter is a journal paper discussing the condition assessment of the affected structure. The above chapter presents findings from the visual inspection and core-based evaluations conducted as a part of the diagnosis. Consequently, the extent of ASR damage along with potential supplementary distress mechanisms are discussed. Moreover, a comparison is conducted between visual inspection and core-based evaluations with respect to their diagnostic character. The fourth chapter is a journal paper proposing a novel semi-empirical model for the prediction of ASR damage in affected structures. The model is developed based upon two previous studies which tackled aspects including aggregate reactivity, alkali content, temperature and relative humidity. The work proposed in chapter four drastically enhances the model's capability to predict damage in real-life structures through its ability to account for directional expansion and various confinement scenarios. Lastly, the fifth chapter presents the conclusions drawn from the above research as well as recommendations for future work. The writing of all chapters in this book along with sample preparation, testing, and data analysis were performed by me. Testing procedures and data analysis conducted in chapter three were carried out with the help of co-author, Andisheh Zahedi (PhD candidate), who contributed to the formulation of the testing matrix and testing plan (i.e., type and number of specimens to be used for each test procedure). Furthermore, conceptualization and methodology applied in chapter four was developed with the help of co-author, Thuc Nguyen (PhD candidate), who provided guidance in the formulation of theoretical concepts to be used in the model and in the detailed analysis of data gathered from literature. Finally, all co-author and supervisor (Dr. Leandro Sanchez), have contributed to the project conceptualization, results analysis and revision of both papers in this book.

1.4   References

[1]   T.E. Stanton, Expansion of concrete through reaction between cement and aggregate, Proc. ASCE. 66 (1940) 1781–1811.

[2]   T. Garcia, M.S. Mirza, Effect of reinforcement on aar expansion in concrete, Proc. Int. Conf. Appl. Codes, Des. Regul. (2005) 271–279.

[3]   B. Fournier, M.A. Bérubé, Alkali-aggregate reaction in concrete: A review of basic concepts and engineering implications, Can. J. Civ. Eng. 27 (2000) 167–191. https://doi.org/10.1139/l99-072.

[4]   A. Mohammadi, E. Ghiasvand, M. Nili, Relation between mechanical properties of concrete and alkali-silica reaction (ASR); a review, Constr. Build. Mater. 258 (2020) 119567. https://doi.org/10.1016/j.conbuildmat.2020.119567.

[5]   C.F. Dunant, K.L. Scrivener, Effects of uniaxial stress on alkali-silica reaction induced expansion of concrete, Cem. Concr. Res. 42 (2012) 567–576. https://doi.org/10.1016/j.cemconres.2011.12.004.

[6]   L.F.M. Sanchez, B. Fournier, D. Mitchell, J. Bastien, Condition assessment of an ASR-affected overpass after nearly 50 years in service, Constr. Build. Mater. 236 (2020) 117554. https://doi.org/10.1016/j.conbuildmat.2019.117554.

[7]   R.B. Figueira, R. Sousa, L. Coelho, M. Azenha, J.M. de Almeida, P.A.S. Jorge, C.J.R. Silva, Alkali-silica reaction in concrete: Mechanisms, mitigation and test methods, Constr. Build. Mater. 222 (2019) 903–931. https://doi.org/10.1016/j.conbuildmat.2019.07.230.

[8]   CSA A23.2-27A, Standard practice to identify degree of alkali-reactivity of aggregates and to identify measures to avoid deletrious expansion in concrete, (2009) 371–384.

[9]   ASTM C1293, Standard test method for determination of length change of concrete due to alkali-silica reaction, Annu. B. ASTM Stand. (2015) 1–7.

[10]   J. Lindgård, Ö. Andiç-Çakir, I. Fernandes, T.F. Rønning, M.D.A. Thomas, Alkali-silica reactions (ASR): Literature review on parameters influencing laboratory performance testing, Cem. Concr. Res. 42 (2012) 223–243. https://doi.org/10.1016/j.cemconres.2011.10.004.

[11]   S. Multon, A. Sellier, Multi-scale analysis of alkali-silica reaction (ASR): Impact of alkali leaching on scale effects affecting expansion tests, Cem. Concr. Res. 81 (2016) 122–133. https://doi.org/10.1016/j.cemconres.2015.12.007.

[12]   N. Sinno, M. Shehata, Understanding the factors affecting expansion due to alkali-silica reaction in concrete, in: 7th Int. Mater. Spec. Conf. 2018, Held as Part Can. Soc. Civ. Eng. Annu. Conf. 2018, 2019: pp. 394–404.

[13]   P. Krivenko, R. Drochytka, A. Gelevera, E. Kavalerova, Mechanism of preventing the alkali-aggregate reaction in alkali activated cement concretes, Cem. Concr. Compos. 45 (2014) 157–165. https://doi.org/10.1016/j.cemconcomp.2013.10.003.

[14]   M.A. Berube, B. Durand, D. Vézina, B. Fournier, Alkali-aggregate reactivity in Québec (Canada), Can. J. Civ. Eng. 27 (2000) 226–245. https://doi.org/10.1139/cjce-27-2-226.

[15]   L.F.M. Sanchez, B. Fournier, M. Jolin, D. Mitchell, J. Bastien, Overall assessment of Alkali-Aggregate Reaction (AAR) in concretes presenting different strengths and incorporating a wide range of reactive aggregate types and natures, Cem. Concr. Res. 93 (2017) 17–31. https://doi.org/10.1016/j.cemconres.2016.12.001.

[16]   L.F.M. Sanchez, T. Drimalas, B. Fournier, D. Mitchell, J. Bastien, Comprehensive damage assessment in concrete affected by different internal swelling reaction (ISR) mechanisms, Cem. Concr. Res. 107 (2018) 284–303. https://doi.org/10.1016/j.cemconres.2018.02.017.

[17]   V. Villeneuve, B. Fournier, J. Duchesne, Determination of the damage in concrete affected by ASR- the damage rating index (DRI), in: Proc. 14th Int. Conf. Alkali-Aggregate React. Concr., 2012.

[18]   T.M. Chrisp, P. Waldron, J.G.M. Wood, Development of a non-destructive test to quantify damage in deteriorated concrete, Mag. Concr. Res. 45 (1993) 247–256.

[19]   T.M. Chrisp, J.G.M. Wood, P. Norris, Towards quantification of microstructural damage in AAR deteriorated concrete, Fract. Concr. Rock, Recent Dev. (1989).

2.1   Background

Over the last decades, a significant number of structures in Canada and around the world have been significantly affected by ASR. From common structures such as roads and bridges to critical infrastructure including dams and nuclear plants, ASR has the potential to greatly impact the level of performance and serviceability in affected concrete. This reaction was first identified in California in the 1940s by Stanton [1]; since then, it has been discovered in structures in over 50 countries around the world [2]. The problem of ASR has raised several concerns regarding the safety and durability of affected structures. Hence, ASR damage has been studied over several decades, with a great deal of the work focused on strategies aimed at the assessment and management of damage to determine appropriate measures for rehabilitation work.

AAR is a chemical reaction that results in the cracking and subsequent expansion of concrete. The cement matrix of concrete generally comprises a considerable concentration of alkalis present in the form of hydroxides (Na^+, K^+ — OH^-) in the concrete pore solution. In the presence of a reactive aggregate (coarse or fine), AAR is triggered between the alkali hydroxides and reactive aggregate, resulting in the formation of a viscous secondary reaction product. In the case of ASR, the source of the reaction in reactive products is unstable mineral phases comprised of poorly crystallized/ uncrystallized silica. The reaction product, known as silica gel or ASR gel, expands upon exposure to moisture ingress, resulting in the development of swelling pressures. Thus, the reaction is directly influenced by three factors: 1) reactive aggregates (i.e., type and nature, size, quantity), 2) alkali hydroxides (i.e., concentration in pore solution, external contribution), and 3) moisture ingress (i.e., internal relative humidity of concrete, external contribution).

2.2   Effects of ASR on concrete

In the context of aging structures affected by ASR, concrete is a durability related issue, since it could lead to the development of important deflections over time in affected members, resulting in reduced serviceability. On a microscopic scale, however, ASR results in the development of significant cracks, which grow in length and size, resulting in the expansion of affected concrete. Thus, understanding the mechanism of ASR distress on the microscopic level is crucial to predicting mechanical behaviour and thus, the structural performance of affected structures.

2.2.1 Microscopic effects of ASR

Since reactive silica is crucial to the development of ASR, concrete distress is greatly dictated by the microstructure of reactive aggregates. In fact, Poole [3] suggested that ASR not dictated by the rock type from which the aggregate was extracted, but rather the mineralogy of the aggregate, since a greater extent of microstructural entropy normally results in more unstable silica minerals [4]. Therefore, amorphous and meta-stable silica crystals are known to be the most reactive mineral phases (i.e., highest expansion levels), followed by micro-crystalline silica [4]. Moreover, internal defects (i.e., microcracks) resulting from geological or industrial processes (i.e., crushing, sieving, etc.) could facilitate damage by providing pathways for the gel to travel into and expand further [4].

It is also important to note that aggregate type and nature (i.e., coarse or fine) plays an important role in crack propagation due to ASR. As shown in Figure 2.1, reactive fine aggregates tend to produce more condensed cracks than reactive coarse aggregates due to their close proximity to each other. Moreover, crushed aggregates normally possess a certain extent of internal defects, resulting in 'sharp' ASR cracks that penetrate through the aggregate [5]. On the other hand, aggregates that are naturally round in geometry, such as sand or gravel, normally exhibit peripheral cracks (i.e., 'onion skin' cracks) [5].

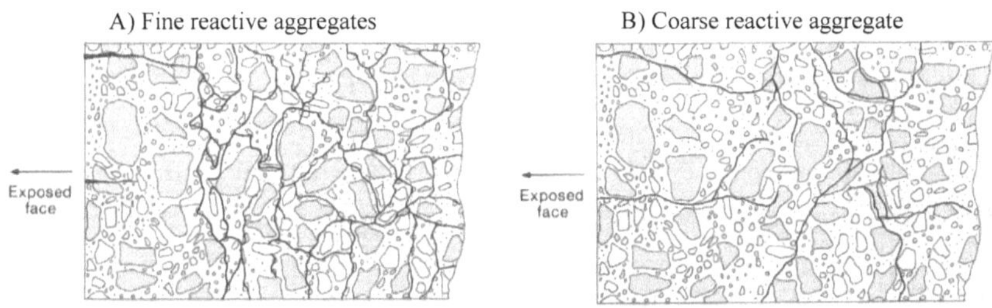

Figure 2.1: Crack propagation for ASR in A) reactive fine aggregates B) reactive coarse aggregates [6].

Sanchez et al. [7] has shown that the above petrographic features are, in fact, characteristic of certain ASR expansion levels. With the aid of microscopic analysis, the authors have developed a model to describe microscopic distress due to ASR as a function of induced expansion (i.e., Figure 2.2). According to the model, in early stages of ASR (i.e., 0.05%), 'type A' (i.e., sharp) cracks may develop in the case of pre-existing defects in the reactive aggregate (i.e., closed cracks due to crushing, sieving, etc.). On the other hand, 'type B' (i.e., onion skin) cracks are more likely to develop in aggregates that do not possess internal microcracks. As expansion reaches moderate levels (i.e., ≈0.12%), type A cracks may begin to extend into the bulk cement paste, whereas type B cracks continue to travel around the aggregate's boundary due to the homogeneous exposure to alkali hydroxides from the pore solution. At this point, expansion continues to increase as existing cracks grow in width and length. At very high expansion levels (i.e., ≥0.30%), type A cracks begin to form a network between reactive aggregates through the bulk cement paste, whereas type B cracks could lead to aggregate debonding, resulting in mechanical implications.

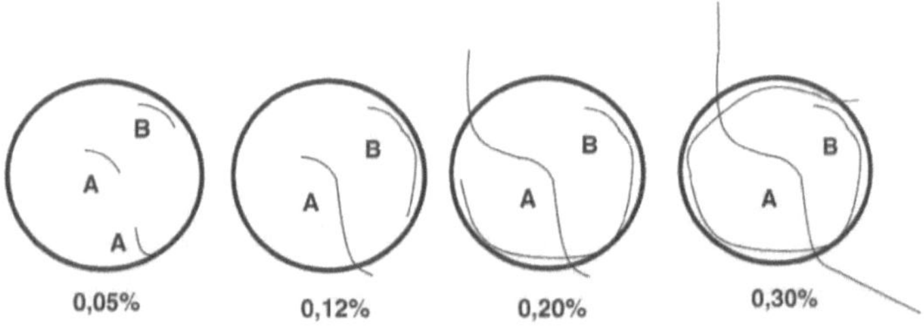

Figure 2.2: AAR damage model proposed by [7].

2.2.2 Mechanical effects of ASR

Comprehending ASR damage on the microstructural scale not only helps to better understand the mechanism of the reaction but can provide insight into the influence of ASR on the mechanical properties of concrete [5,8–13]. In fact, the inevitable presence of internal defects is a key reason behind the different responses exhibited by sound concrete under different confinement scenarios (i.e., brittle response under uniaxial loading, and ductile response in confined conditions). In the presence of damage mechanisms, such distress becomes more complex to predict, due to the intrinsic heterogeneous property of concrete. Researchers have shown that reductions in tensile strength and the modulus of elasticity due to ASR distress tend to be sharper and greater than reduction in compressive strength [10,14–16]. Moreover, compressive strength reduction only begins during later stages of the reaction (i.e., higher expansion levels – 0.30% or higher) [16]. This is essentially due to the fact that concrete the concrete modulus of elasticity is dictated by the stiffness of coarse aggregates, whereas the mode of concrete failure in compression is governed by the bulk cement paste.

2.3 Factors influencing ASR distress in affected structures

In a controlled laboratory setting, ASR expansion and distress are relatively straightforward to predict. However, structural members in the field are almost always subjected to unique conditions that are difficult to recreate in laboratory settings yet are crucial to the development of ASR. Among those, exposure conditions and confinement are perhaps the most important.

2.3.1 Exposure conditions

Although it is quite well-documented that moisture plays an important role in ASR expansion [17–19], different views exist among the research community as to its role in the reaction [20–23]. It is generally agreed upon that an internal RH greater than 80% is required for ASR-induced expansion to take place. In the event that RH drops below the above value, Multon et al. [22] suggests that the reaction may resume once placed in a humid environment once again. On the other hand, according to Larive [20], once the initial moisture present during gel production is consumed, the addition of water will not result in further expansion [24]. Despite the above threshold RH requirement (i.e., RH > 80%), Poyet et al. [21] showed that expansion occurs in a laboratory specimen stored at 60°C and 59% RH, suggesting that expansion also depends on temperature, as is the case with any chemical reaction. Nevertheless, according to Larive [25], temperature tends to affect reaction kinetics (i.e., how fast ultimate expansion is reached), whereas RH influences the ultimate expansion. The above was demonstrated by both Ulm et al. [26] and Comi et al [27], who mentioned that ASR and its effects are dependent on the coupled problem of reaction kinetics and heat diffusion [26].

2.3.2 Confinement conditions

Research efforts aimed at identifying the effect of confinement (internally and externally) have led to numerous interpretations within the research community, many of which are conflicting [28–32]. Nevertheless, it is generally agreed upon that the existence of restraints, in the form of reinforcement or uniaxial/biaxial confining stresses, may lead to the reduction of "free expansion" deformations resulting from ASR [28,29,32]. One of the most significant reports outlining the impact of restraints on ASR development was published by the Institution of Structural Engineers (ISE) [28] in 1992, compiling expansion results from a variety of researchers. Findings in the above report suggest that restraints may exacerbate the physical implications of ASR, including surface microcracking and induced compressive stresses in concrete and tensile stresses in reinforcement. Furthermore, confinement tends to restrict expansion to the direction parallel to reinforcing steel. Figure 2.3 displays the trend of restrained/free expansion versus the reinforcement ratio, indicating a sharp initial drop with rising reinforcement content, followed by a steady leveling off. The above indicates that a small amount of reinforcement may greatly reduce ASR expansion. In fact, experimental results reported by Hobbs [17] demonstrate that reinforcement as little as 0.5% may result in the drastic reduction of expansion strains (i.e., up to around 55%). Moreover, a reduction of up to 90% was achieved through a reinforcement ratio of 4% by the same author [17]. One may also note that induced compressive stress in affected concrete tends to rise with greater expansion. According to the ISE [28], it is evident that the maximum compressive stress induced in concrete is limited to 4 MPa, suggesting that the yielding of reinforcement is likely in concretes showcasing adequate expansive strains. This phenomenon was also reported in works by Swamy [33] (1987), Al Asali [34] (1989) and Wood [14], who demonstrated that stresses induced through AAR may be sufficient to stimulate yielding of steel reinforcement in a 0.9% reinforced concrete. Nevertheless, this structural implication may be counteracted by increasing the amount of reinforcement, as this leads to a reduction in induced net tensile stresses [28]. In 2005, Garcia carried out a study to examine the effect of reinforcement on the free expansion of reactive Spratt aggregate, in which reductions of 40% and 60% were achieved with reinforcement ratios of 4% and 7%, respectively [35]. Later in 2007, Smaoui developed a similar relationship (i.e., restrained/free expansion vs reinforcement ratio) for New Mexico gravel and Québec City limestone [31]. In terms of the influence of the reactive aggregate involved, Smaoui also importantly concluded that the relationship between restrained and free expansion for reinforced members was quite similar for coarse and fine aggregates, despite varying expansion rates and ultimate expansion [31].

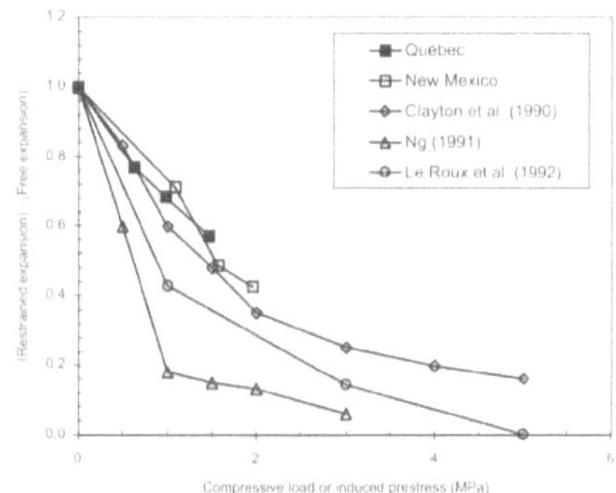

Figure 2.3: Effect of reinforcement and stress confinement [31].

The orientation of reinforcement is critical to effective confinement of AAR expansion, as cracking tends to be restricted in the direction parallel to rebars. However, the above condition requires a homogeneous distribution of reinforcement through the cross section of the affected member [28]. Figure 2.4 demonstrates the macroscopic pattern exhibited by surface cracks due to the aforementioned effect yet highlights that reinforcement must be uniformly distributed to achieve that effect. On the other hand, researchers have shown that transversal reinforcement is not nearly as effective at reducing expansive strains. The ISE (1992) suggests that transversal strains (i.e., along the direction of stirrups) tend to be significantly more pronounced than axial strains (i.e., along the direction of longitudinal reinforcement) [28]. Similarly, Larive [25] (1998) advocated that uniaxial loading tends to reduce expansion along that direction yet compensates for the reduction through an increase in strains along supplementary directions, such as to maintain a constant volumetric expansion. Jones and Clark [36] pursued a similar proposition, suggesting that neither internal nor external confinement (i.e., due to stirrups or uniaxial compressive loading, respectively) notably supress expansion perpendicular to the direction of reinforcement. This apparent ineffectiveness of stirrups, however, was explained in 2006 by Multon and Toutlemonde [37], who confirmed, through a comprehensive experimental program, the phenomenon dictating the directional transfer of expansive strains depending on the direction and magnitude of confinement. In their study, Multon and Toutlemonde [37] monitored the effects of different axial and radial confinement on the directional and volumetric expansion of cylindrical concrete specimens. Interestingly, through the expansion monitoring of reinforced slabs (610 x 750 x 4500 mm), Allard et al. [30] (2018) suggested that longitudinal reinforcement may lead to somewhat effective confinement (i.e., expansion limited to 0.009%) along the perpendicular direction for non-highly reactive aggregates, yet the above greatly depends on the reinforcement ratio. The same authors however, illustrated that stirrups were, in fact, ineffective at deterring ASR expansion strains. Nevertheless, this observation supports the claim that deformation may have been transferred in the direction of lower restraint.

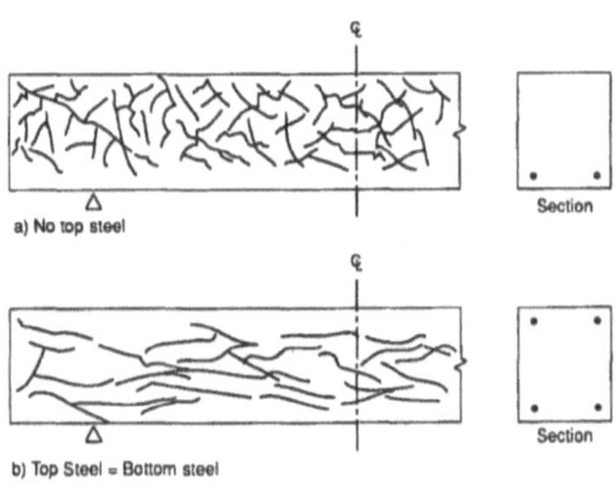

Figure 2.4: Effect of reinforcement on crack orientation [28].

In addition to the confinement effect of steel reinforcement, external compressive stress can be quite effective at reducing expansive strains consequential to ASR development. In addition to confirming the deformation transfer phenomenon (i.e., expansion tends to be more drastic in the less confined direction), Multon & Toutlemonde [37] suggested that volumetric expansion remains more or less constant, irrespective of the stress state. Similar studies were carried out by Gautam et al. [38] (2017) and Liaudat et al. [39] (2018), wherein ASR affected cubical concrete specimens were examined under different stress states. A model describing volumetric confinement as a function of applied uniaxial stress was proposed, indicating a sharp decline in axial expansion between 1 and 3 MPa, followed by a plateau, beyond which the confinement level remains constant. The aforementioned model, proposed by Gautam et al. [38] (2017), seems somewhat reasonably representative of the performance of field elements that tend to be subjected to compressive stresses less than 10 MPa. Contrary to Multon & Toutlemonde [37], however,

Gautam et al. [38] and Liaudat et al. [39] suggested that volumetric expansion tends to decrease with the increased application of volumetric compressive stress.

2.4 Tools for the appraisal of ASR in structures

2.4.1 Non-destructive methods

2.4.1.1 Visual inspection

The first step in appraising any structure speculated to be damaged is by assessing the current conditions of elements through visual inspection. During this process, typical signs of ASR distress are reported including cracking, spalling, gel exudations, and so on. Information relating to cracks is often reported in terms of average crack widths and patterns. Furthermore, a couple of crack mapping techniques have been developed as a means of quantifying surface damage.

One method reported in a publication by the Institution of Structural Engineers [28] involves the summation of crack widths along five parallel lines constructed on the concrete surface, and then dividing by the length of lines drawn. Another method proposed by Godart et al. [40] involves the construction of a grid to average the crack widths along perpendicular lines and the 45° diagonal lines intercepting them. The aforementioned is known as the cracking index (CI) method. The CI method appears to be more representative of ASR damage since it accounts for the random pattern of cracks (i.e., map cracking) associated with ASR.

A)   B)

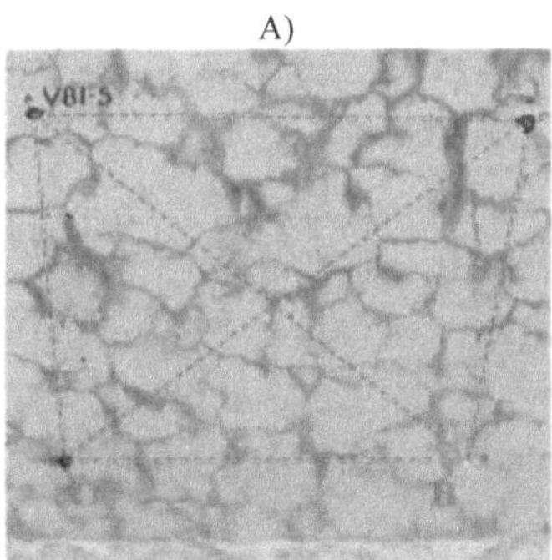
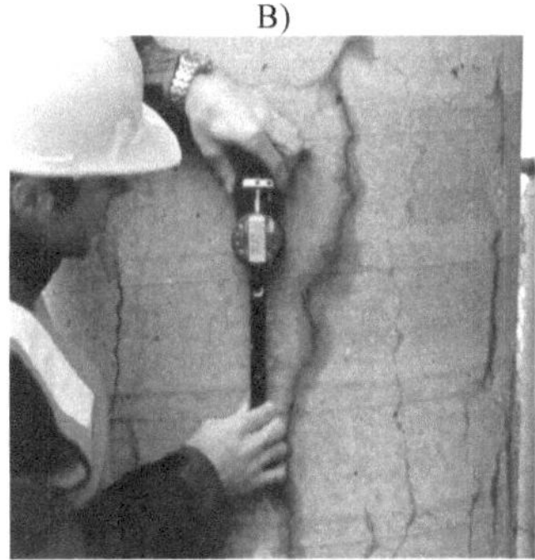

Figure 2.5: A) Cracking index grid B) expansion measurement [41].

Nevertheless, crack mapping techniques have achieved conflicting judgement relating to their effectiveness among authors. Smaoui et al. [42] reported lower predicted strains than actual strains discovered in laboratory specimens. Specimens exposed to the outdoor environment, however, displayed varying results depending on the level of exposure of a particular specimen [43].

2.4.2   Evaluation of core specimens

A quite useful method of appraising aging structures is the evaluation of core specimens extracted from the deteriorating structural element. The core drilling process is relatively unharmful to the structure itself, so long as a structural analysis is prepared to identify the optimal drilling locations such as to avoid damage to internal reinforcements or serviceability of the element. Nevertheless, a thorough condition assessment of the affected structure may be conducted using a limited number of specimens. In fact, certain test procedures may be classified as non-destructive (i.e., the Stiffness Damage Test – SDT), warranting investigation using additional techniques. Therefore, through comprehensive planning, one could minimize the number of cores necessary for a condition assessment.

Diagnostic core evaluation methods for the condition assessment of ASR affected structures may be classified into two categories: 1) microscopic tools and 2) mechanical tools. The following section discusses the aforementioned testing protocols and their objectives. Recommendations will also be provided regarding the use of one core specimen for conducting multiple test procedures.

2.4.2.1 Petrographic examination

The microscopic examination of concrete thin sections can yield quite useful information regarding the mineralogy and condition of affected concrete. The analysis is carried out, according to ASTM C856 standards [44], using a stereoscope under varying lighting conditions and magnifications, typically accompanied by other methods (i.e., scanning electron microscopy – SEM, infrared spectroscopy, X-ray diffraction or others). As a result, petrographers may identify the composition (i.e., rock type and mineral phases) of aggregates used (both coarse and fine) as well as petrographic features including microcracks in the aggregate particle or cement paste, cracks in the ITZ, reaction products and so forth, that are associated with a particular distress mechanism (i.e., ASR, DEF, corrosion, sulphate attack) [45].

2.4.2.2 Damage Rating Index

The Damage Rating Index is a semi-quantitative petrographic tool used to quantify the level of damage associated with internal swelling reactions (ISR). Contrary to petrographic examination, however, this method cannot be used to confirm the presence of distress, but rather, to determine the extent of damage. Grattan Bellew and coworkers [46] first developed this microscopic technique for the evaluation of ASR damage; Villeneuve & Fournier [47] later proposed a modification by modifying weighting factors, whose purpose is to account for the relative importance of petrographic features towards their contribution to a particular distress mechanism (e.g., ASR). Over the years, studies have shown that DRI is a very powerful tool for the appraisal of AAR damage in affected concrete specimens.

2.4.2.3 Compressive strength test

The compressive strength of concrete specimens, determined through CSA 23.2-9C [48] is quite a significant measure of concrete performance and serviceability, particularly in aging structures. However, research has suggested that compressive strength is not a good indicator of AAR damage/progress, particularly since it is sensitive only to greater levels of damage (i.e., expansion). Nevertheless, in the case of core-based evaluations, the CS test is important for evaluating overall concrete condition as it provides an indication of the mechanical performance of field concrete. Thus, it is quite significant in the context of ensuring structural safety first. Secondly, the CS is also important since it provides a reference point for sound concrete when conducting the SDT.

2.4.2.4 Stiffness Damage Test

The Stiffness Damage Test (SDT) was first developed by Walsh [49] for the testing of rock specimens under cyclic loading. Walsh discovered a good correlation between resulting crack patterns and the stress/strain relationship [49]. Based on the above findings, Crouch [14] proposed a new procedure involving cyclic loading of cylindrical concrete specimens in compression. Crisp [50,51] later suggested the use of the SDT for the evaluation of ASR distress in concrete, where samples are loaded at a rate of 0.10MPa/s with an initial fixed load of 5.5MPa. In order to optimize the procedure for diagnosis, Smaoui et al. [43,52] proposed that the initial fixed load be increased to 10MPa.

The work of Smaoui et al. was expanded on by Sanchez et al. [10,53,54], who carried out the SDT on concrete mixtures of various strengths (i.e., 25, 35, and 45MPa), containing a variety of reactive aggregates. Based on the above, the authors proposed a modification by altering the initial fixed loading to 40% of the compressive strength of the tested specimen [10,53,54]. Furthermore, in addition to determining the modulus of elasticity of tested concrete, promising output parameters for the diagnosis of ASR distress were identified: The Stiffness Damage Index (SDI) and the Plasticity Deformation Index (PDI). The latter are accurately able to express damage as a ratio of dissipated energy/total energy and plastic deformation/total deformation, respectively. Moreover, the non-linearity index (NLI) was deemed a quite interesting parameter for consideration as a complimentary output parameter, as it is able to forecast the level of damage as well as the pattern/direction of main cracks resulting from ASR [10,50,51,53,54].

2.5   Prognosis of ASR-affected structures

The condition assessment of ASR is quite a significant step in evaluating the existing damage degree in affected concrete. However, to implement an efficient management protocol in aging structures, it is important to understand how the reaction progresses over time (i.e., time-dependent expansion curve). The above considerations are important for forecasting the impact on physical integrity and mechanical properties due to ASR distress, and thus the overall performance of affected concrete.

2.5.1   Expansion testing

Accelerated expansion testing in the laboratory is perhaps one of the most commonly used tools for the prediction of ASR expansion [55–59]. Among the developed protocols for expansion testing, two in particular have become widely applied for the prediction of ASR. The accelerated mortar bar test (AMBT - ASTM C1260 and CSA A23.2-25A), first proposed by Oberholster and Davis in 1986 [55], is perhaps the most commonly used expansion test. The concrete prism test (CPT – ASTM C1293 and CSA A23.2-14A), on the other hand, is viewed by most researchers as the most reliable test procedure for evaluating expansion. Both tests are conducted in accelerated conditions (i.e., high temperature and RH) to expedite the expansion of test specimens. Despite the considerably shorter test duration with the AMBT (i.e., 14 days compared to one year for CPT), the CPT is able to detect alkali-silica reactivity, alkali-carbonate reactivity, and the reactivity of mixtures containing supplementary cementitious materials (SCMs). Furthermore, several instances have been recorded with the AMBT where the test misjudges non-reactive aggregates as reactive and is unable to detect the reactivity of aggregates considered reactive as per the CPT [55].

Expansion tests have even been used as a prognostic tool for affected structures. Bérube et al. [57] proposed the "residual expansion test" designed to predict the potential for remaining expansion of cores extracted from affected elements. The above procedure may be carried out by testing core specimens in accelerated conditions (i.e., 38°C and 100% RH). Alternatively, expansion testing may be carried out in a 1M NaOH solution at 38°C, or preferably, according to the CPT procedure for reactive coarse aggregates extracted from core specimens.

2.5.2   ASR Models

Despite being quite reliable in identifying aggregate reactivity (i.e., ultimate expansion), expansion test procedures (e.g., CPT and AMBT) are often limited by laboratory equipment. Therefore, interpreting such results directly in the context of predicted expansion in structural members is premature, since many factors (mentioned in section 2.3) remain unaccounted for. This gap has been addressed by researchers through numerous modeling approaches which have been developed to describe the reaction mechanism and effects at different levels. One may classify models according to three levels: 1) macroscopic: predicting behaviour on a structural level, 2) mesoscopic: models predicting behaviour on the aggregate level, and 3) microscopic: predicting behaviour on the level of reaction products [60].

The kinetic property of expansion (i.e., time-dependency) is one of the most crucial drawbacks of using expansion tests to predict structural damage, since the accelerated nature of such tests (i.e., high temperature and relative humidity) neglects the timeline of expansion. Thus, one of the most significant numerical models that have been developed is the approach proposed by Larive [25] based on a thorough experimental campaign using more than 600 test specimens.

According to Larive [25], ASR expansion may be characterized by two parameters relating to reaction kinetics (i.e., latency - τ_L and characteristic - τ_C times) and a parameter for the ultimate asymptotic expansion (Eq. 2.1). It is worth mentioning that reaction kinetics are influenced by temperature, whereas the ultimate expansion value is affected by relative humidity. Thus, each set of parameters is unique for a particular reactive aggregate at a given temperature. Figure 2.6 demonstrates a sample expansion curve, annotated based on Larive's proposed numerical model. According to the author, the reaction may be dissected into four stages, describing the reaction mechanism. The first stage is characterized by the accumulation of reaction product (i.e., ASR gel) within reactive aggregates, resulting in negligible expansion. Afterwards, moisture ingress leads to the swelling of silica gel, resulting in the exponential rise in the second stage of the expansion curve. Past the inflection point, major crack formation takes place in the affected concrete, resulting in the generation of space to accommodate the swelling product. Therefore, the curve changes shape from convex to concave as shown in Figure 2.6. Finally, the depletion of alkali hydroxides in the system leads to the inevitable plateau of the expansion curve.

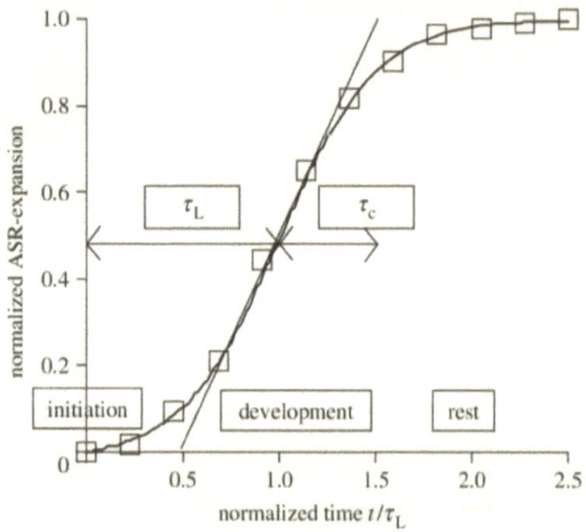

Figure 2.6: Expansion curve based on Larive's equation [62].

$$\varepsilon(t, \theta) = \frac{1 - e^{-\frac{t}{\tau_C(\theta)}}}{1 + e^{-\frac{t-\tau_L(\theta)}{\tau_C(\theta)}}} \times \varepsilon^{\infty} \tag{2.1}$$

Recently, De Grazia et al. [61] calibrated the model using experimental data from a wide range of concrete mixtures, which provides the equation with a physicochemical meaning, such that may be used to model the expansion of laboratory specimens. Consequently, [61] proposed a novel semi-empirical model based on [25] which accounts for the type and nature of reactive aggregate, alkali content, temperature, and relative humidity.

Capturing the effect of exposure conditions (i.e., temperature and relative humidity) on expansion is essential for interpreting laboratory expansion in the context of expansion in the field. This was shown by Ulm et al. [26], who expanded on Larive's model by addressing the coupled issue of reaction kinetics and heat diffusion, and thus developed Eq. 2.2 and 2.3 , where U_C and U_L represent activation energy as per the Arrhenius concept [26]. Moreover, Capra developed a model (Eq. 2.4) for the calculation of normalized ultimate expansion (i.e., as a fraction of expansion at a relative humidity of 100%) as a function of internal relative humidity (h) [63]. Finally, according to Kawabata et al. [64], for a given concrete mixture at a given temperature, there exists an expansion with unique kinetics (i.e., characteristic and latency times) that correspond only to that temperature, known as a 'master curve'. The author proposed a procedure, based on the above, for the transition from one master curve to another, while accounting for the timescale of the reaction (Figure 2.7).

$$\tau_C(\theta) = \tau_C(\theta_0)\exp\left(U_C\left(\frac{1}{\theta} - \frac{1}{\theta_0}\right)\right) \tag{2.2}$$

$$\tau_L(\theta) = \tau_L(\theta_0)\exp\left(U_L\left(\frac{1}{\theta} - \frac{1}{\theta_0}\right)\right) \tag{2.3}$$

$$\frac{\varepsilon_{RH}}{\varepsilon_{RH=100}} = h^8 \tag{2.4}$$

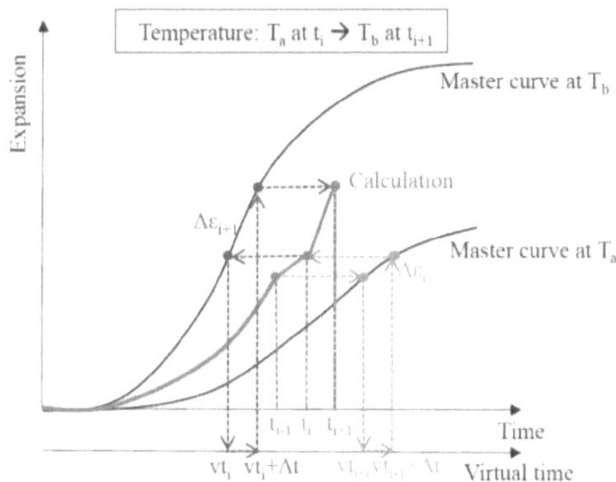

Figure 2.7: Calculation procedure for correlation between laboratory and field expansion [64].

2.6   Summary

Several procedures and protocols have been developed over the years for the appraisal of ASR-affected structures. More recently, the multi-level assessment (i.e., micro-mechanical coupling) as per Sanchez et al [5,9] has demonstrated capability to diagnose ASR damage in affected concrete through the use of DRI and SDT procedures. Nevertheless, applications to structures in the field have been quite limited so far; this work aims to address this issue by conducting the multi-level assessment on an ASR-affected aging concrete structure.

Moreover, current protocols for the prognosis of ASR using core-based evaluation (e.g., residual expansion test) are often too time consuming and only provide a "reference" for the predicted expansion in the field, since significant factors are not accounted for. Anisotropy, perhaps the most important factor, can dramatically reduce or increase expansion in a given direction, resulting in potential implications to concrete properties, and thus is key to detecting structural issues that may arise due to ASR. Replicating such effects in a laboratory environment is quite difficult and often impractical. Furthermore, literature has revealed that no models currently exist to directly simulate expansion in structures considering the aforementioned influences. However, existing semi-empirical and analytical models could potentially be enhanced to achieve this objective. Larive's equation, for instance, is able to express time-dependent expansion as a function of ultimate expansion and reaction kinetics [25]. Recent studies have successfully attributed a physical meaning to the equation's parameters, enabling the modeling of expansion test specimens in a laboratory setting [61]. Furthermore, several researchers have investigated factors influencing anisotropy (i.e., specimen geometry and confinement conditions) in ASR affected concrete. Through a comprehensive statistical analysis of existing experimental data, a relationship can be developed to better understand the effect of such factors and to incorporate them into Larive's equation to accurately model expansion in existing structures. Consequently, the aim of this work will be to conduct a full assessment (i.e., diagnosis and prognosis) of an existing ASR-affected structure through the use of the multi-level assessment protocol and a novel ASR expansion model to determine the existing level of damage and forecast future ASR development over the service life of the considered structure.

2.7   References

[1]   T.E. Stanton, Expansion of concrete through reaction between cement and aggregate, Proc. ASCE. 66 (1940) 1781–1811.

[2]   M.A. Berube, B. Durand, D. Vézina, B. Fournier, Alkali-aggregate reactivity in Québec (Canada), Can. J. Civ. Eng. 27 (2000) 226–245. https://doi.org/10.1139/cjce-27-2-226.

[3]   A.B. Poole, Introduction to alkali-aggregate reaction in concrete, in: Alkali-Silica React. Concr., CRC Press, 1991: pp. 17–45.

[4]   F. Rajabipour, E. Giannini, C. Dunant, J.H. Ideker, M.D.A. Thomas, Alkali-silica reaction: Current understanding of the reaction mechanisms and the knowledge gaps, Cem. Concr. Res. 76 (2015) 130–146. https://doi.org/10.1016/j.cemconres.2015.05.024.

[5]   L.F.M. Sanchez, T. Drimalas, B. Fournier, D. Mitchell, J. Bastien, Comprehensive damage assessment in concrete affected by different internal swelling reaction (ISR) mechanisms, Cem. Concr. Res. 107 (2018) 284–303. https://doi.org/10.1016/j.cemconres.2018.02.017.

[6]   D. Palmer, The diagnosis of alkali–silica reaction. Report of a working party, Br. Cem. Assoc. Wexham Springs, Slough. (1992).

[7]   L.F.M. Sanchez, B. Fournier, M. Jolin, J. Duchesne, Reliable quantification of AAR damage through assessment of the Damage Rating Index (DRI), Cem. Concr. Res. 67 (2015) 74–92. https://doi.org/10.1016/j.cemconres.2014.08.002.

[8]   A. Mohammadi, E. Ghiasvand, M. Nili, Relation between mechanical properties of concrete and alkali-silica reaction (ASR); a review, Constr. Build. Mater. 258 (2020) 119567. https://doi.org/10.1016/j.conbuildmat.2020.119567.

[9]   L.F.M. Sanchez, B. Fournier, M. Jolin, D. Mitchell, J. Bastien, Overall assessment of Alkali-Aggregate Reaction (AAR) in concretes presenting different strengths and incorporating a wide range of reactive aggregate types and natures, Cem. Concr. Res. 93 (2017) 17–31. https://doi.org/10.1016/j.cemconres.2016.12.001.

[10]   L.F.M. Sanchez, Contribution to the assessment of damage in aging concrete infrastructures affected by alkali-aggregate reaction, PhD., (2014) 377.

[11]   T. Siemes, N. Han, J. Visser, Unexpectedly low tensile strength in concrete structures,

Heron. (2002).

[12] I. Yurtdas, D. Chen, D.W. Hu, J.F. Shao, Influence of alkali silica reaction (ASR) on mechanical properties of mortar, Constr. Build. Mater. 47 (2013) 165–174.

[13] Y. Kubo, M. Nakata, Effect of reactive aggregate on mechanical properties of concrete affected by alkali-silica reaction., in: 14th ICAAR - Int. Conf. Alkali-Aggregate React. Concr., Austin (Texas), 2012.

[14] R.S. Crouch, J.G.M. Wood, Damage evolution in AAR affected concretes, Eng. Fract. Mech. 35 (1990) 211–218. https://doi.org/10.1016/0013-7944(90)90199-Q.

[15] N. Smaoui, M.-A. Bérubé, B. Fournier, B. Bissonnette, Influence of Specimen Geometry, Orientation of Casting Plane, and Mode of Concrete Consolidation on Expansion Due to ASR, 2004.

[16] Y. Kubo, M. Nakata, Effect of reactive aggregate on mechanical properties of concrete affected by alkali–silica reaction, in: 14th Int. Conf. Alkali–Aggregate React. (ICAAR). Texas, USA, 2012.

[17] D.W. Hobbs, Alkali-silica reaction in concrete, Thomas Telford Publishing, 1988.

[18] H. Olafsson, The effect of relative humidity and temperature on alkali expansion of mortar bars, in: Proc., 7th Int. Conf. Alkali Aggreg. React. Concr., 1986: pp. 461–465.

[19] T. Kurihara, K. Katawaki, Effects of moisture control and inhibition on alkali silica reaction, in: Proc., 8th Int. Conf. Alkali Aggreg. React. Concr., 1989: pp. 629–634.

[20] C. Larive, A. Laplaud, O. Coussy, The role of water in alkali-silica reaction, Proc. 11th ICAAR. 1 (2000) 8.

[21] S. Poyet, A. Sellier, B. Capra, G. Thèvenin-Foray, J.-M. Torrenti, H. Tournier-Cognon, E. Bourdarot, Influence of Water on Alkali-Silica Reaction: Experimental Study and Numerical Simulations, J. Mater. Civ. Eng. 18 (2006) 588–596. https://doi.org/10.1061/(asce)0899-1561(2006)18:4(588).

[22] S. Multon, F. Toutlemonde, Effect of moisture conditions and transfers on alkali silica reaction damaged structures, Cem. Concr. Res. 40 (2010) 924–934. https://doi.org/10.1016/j.cemconres.2010.01.011.

[23] D. Stark, The moisture condition of field concrete exhibiting alkali-silica reactivity, Spec. Publ. 126 (1991) 973–988.

[24] D.G.R. Bonnell, Studies in gels IV.—The swelling of silica gel, Trans. Faraday Soc. 29 (1933) 1217–1220.

[25] C. Larive, Apports combinés de l'expérimentation et de la modélisation à la compréhension de l'Alcali-réaction et de ses effets mécaniques, in: 1998.

[26] F.-J. Ulm, O. Coussy, L. Kefei, C. Larive, Thermo-Chemo-Mechanics of ASR Expansion in Concrete Structures, J. Eng. Mech. 126 (2000) 233–242.

[27] C. Comi, B. Kirchmayr, R. Pignatelli, Two-phase damage modeling of concrete affected by alkali-silica reaction under variable temperature and humidity conditions, Int. J. Solids Struct. 49 (2012) 3367–3380. https://doi.org/10.1016/j.ijsolstr.2012.07.015.

[28] The Institution of Structural Engineers, Structural effects of alkali-silica reaction: Technical guidance on the appraisal of existing structures, 1992.

[29] H. Aryan, B. Gencturk, M. Hanifehzadeh, J. Wei, ASR Degradation and Expansion of Plain and Reinforced Concrete, Struct. Congr. 2020 - Sel. Pap. from Struct. Congr. 2020. (2020) 303–315. https://doi.org/10.1061/9780784482896.029.

[30] A. Allard, S. Bilodeau, F. Pissot, B. Fournier, J. Bastien, B. Bissonnette, Expansive behavior of thick concrete slabs affected by alkali-silica reaction (ASR), Constr. Build. Mater. 171 (2018) 421–436. https://doi.org/10.1016/j.conbuildmat.2018.03.159.

[31] N. Smaoui, B. Bissonnette, M.A. Bérubé, B. Fournier, Stresses induced by alkali-silica reactivity in prototypes of reinforced concrete columns incorporating various types of reactive aggregates, Can. J. Civ. Eng. 34 (2007) 1554–1566. https://doi.org/10.1139/L07-063.

[32] A. Zahedi, C. Trottier, L.F.M. Sanchez, M. Noël, Microscopic assessment of ASR-affected concrete under confinement conditions, Cem. Concr. Res. 145 (2021) 106456. https://doi.org/10.1016/j.cemconres.2021.106456.

[33] R.N. Swamy, M.M. Al-Asali, Engineering Properties of Concrete Affected By Alkali-Silica Reaction., ACI Mater. J. 85 (1988) 367–374. https://doi.org/10.14359/2288.

[34] R. Narayan Swamy, M.M. Al-Asali, Effect of alkali-silica reaction on the structural behavior of reinforced concrete beams, ACI Struct. J. 86 (1989) 451–459. https://doi.org/10.14359/2961.

[35] T. Garcia, M.S. Mirza, Effect of reinforcement on aar expansion in concrete, Proc. Int. Conf. Appl. Codes, Des. Regul. (2005) 271–279.

[36] A.E.K. Jones, L.A. Clark, The practicalities and theory of using crack width summation to estimate ASR expansion, Proc. Inst. Civ. Eng. Struct. Build. 104 (1994) 183–192. https://doi.org/10.1680/istbu.1994.26327.

[37] S. Multon, F. Toutlemonde, Effect of applied stresses on alkali-silica reaction-induced expansions, Cem. Concr. Res. 36 (2006) 912–920. https://doi.org/10.1016/j.cemconres.2005.11.012.

[38] B.P. Gautam, D.K. Panesar, S.A. Sheikh, F.J. Vecchio, Multiaxial expansion-stress relationship for alkali silica reaction-affected concrete, ACI Mater. J. 114 (2017) 171–184. https://doi.org/10.14359/51689490.

[39] J. Liaudat, I. Carol, C.M. López, V.E. Saouma, ASR expansions in concrete under triaxial confinement, Cem. Concr. Compos. 86 (2018) 160–170. https://doi.org/10.1016/j.cemconcomp.2017.10.010.

[40] B. Godart, P. Fasseu, M. Michel, Diagnosis and monitoring of concrete bridges damaged by AAR in Northern France, in: 9th Int. Conf. Alkali-Aggregate React. Concr. July 1992, London, Vol. 1, 1992.

[41] B. Fournier, M.A. Berube, K.J. Folliard, M. Thomas, Report on the Diagnosis, Prognosis, and Mitigation of Alkali-Silica Rection (ASR) in Transportation Structures, Austin, 2010.

[42] N. Smaoui, B. Fournier, M.-A. Bérubé, B. Bissonnette, B. Durand, Evaluation of the expansion attained to date by concrete affected by alkali silica reaction. Part II: Application to nonreinforced concrete specimens exposed outside, Can. J. Civ. Eng. 31 (2004) 997–1011. https://doi.org/10.1139/l04-074.

[43] M.A. Bérubé, N. Smaoui, B. Fournier, B. Bissonnette, B. Durand, Evaluation of the expansion attained to date by concrete affected by ASR - Part III: Application to existing structures, Can. J. Civ. Eng. 32 (2005) 463–479.

[44] ASTM C856-20, Standard Practice for Petrographic Examination of Hardened Concrete, Annu. B. ASTM Stand. i (2004) 1–17. https://doi.org/10.1520/C0856.

[45] B. Fournier, M.A. Bérubé, Alkali-aggregate reaction in concrete: A review of basic concepts and engineering implications, Can. J. Civ. Eng. 27 (2000) 167–191. https://doi.org/10.1139/l99-072.

[46] P.E. Grattan-Bellew, L.D. Mitchell, Quantitative petrographic analysis of concrete: the damage rating index (DRI) method, a review, (2006).

[47] V. Villeneuve, B. Fournier, J. Duchesne, Determination of the damage in concrete affected by ASR- the damage rating index (DRI), in: Proc. 14th Int. Conf. Alkali-Aggregate React. Concr., 2012.

[48] Standard counsil of Canada, CSA23.2-9C: Compressive strength of cylindrical concrete specimens, in: CSA A23.119/CSA A23.219 Natl. Stand. Canada, 2019: pp. 733–748.

[49] J.B. Walsh, The effect of cracks on the uniaxial elastic compression of rocks, J. Geophys. Res. 70 (1965) 399–411.

[50] T.M. Chrisp, P. Waldron, J.G.M. Wood, Development of a non-destructive test to quantify damage in deteriorated concrete, Mag. Concr. Res. 45 (1993) 247–256.

[51] T.M. Chrisp, J.G.M. Wood, P. Norris, Towards quantification of microstructural damage in AAR deteriorated concrete, Fract. Concr. Rock, Recent Dev. (1989).

[52] N. Smaoui, M.A. Bérubé, B. Fournier, B. Bissonnette, B. Durand, Evaluation of the expansion attained to date by concrete affected by alkali-silica reaction. Part I: Experimental study, Can. J. Civ. Eng. (2004). https://doi.org/10.1139/L04-051.

[53] L.F.M. Sanchez, B. Fournier, M. Jolin, J. Bastien, Evaluation of the stiffness damage test (SDT) as a tool for assessing damage in concrete due to ASR: Test loading and output responses for concretes incorporating fine or coarse reactive aggregates, Cem. Concr. Res. 56 (2014) 213–229. https://doi.org/10.1016/j.cemconres.2013.11.003.

[54] L.F.M. Sanchez, B. Fournier, M. Jolin, J. Bastien, Evaluation of the Stiffness Damage Test (SDT) as a tool for assessing damage in concrete due to alkali-silica reaction (ASR): Input parameters and variability of the test responses, Constr. Build. Mater. 77 (2015) 20–32.

https://doi.org/10.1016/j.conbuildmat.2014.11.071.

[55] D. Lu, B. Fournier, P.E. Grattan-Bellew, Evaluation of accelerated test methods for determining alkali-silica reactivity of concrete aggregates, Cem. Concr. Compos. 28 (2006) 546–554. https://doi.org/10.1016/j.cemconcomp.2006.03.001.

[56] M. Thomas, B. Fournier, K. Folliard, J. Ideker, M. Shehata, Test methods for evaluating preventive measures for controlling expansion due to alkali-silica reaction in concrete, Cem. Concr. Res. 36 (2006) 1842–1856. https://doi.org/10.1016/j.cemconres.2006.01.014.

[57] M.A. Berube, J. Frenette, A. Pedneault, M. Rivest, Laboratory assessment of the potential rate of ASR expansion of field concrete, Cem. Concr. Aggregates. 24 (2002) 13–19. https://doi.org/10.1520/cca10486j.

[58] J.H. Ideker, B.L. East, K.J. Folliard, M.D.A. Thomas, B. Fournier, The current state of the accelerated concrete prism test, Cem. Concr. Res. 40 (2010) 550–555. https://doi.org/10.1016/j.cemconres.2009.08.030.

[59] X. Mo, B. Fournier, Investigation of structural properties associated with alkali-silica reaction by means of macro- and micro-structural analysis, Mater. Charact. 58 (2007) 179–189. https://doi.org/10.1016/j.matchar.2006.04.018.

[60] R. Esposito, M.A.N. Hendriks, Literature review of modelling approaches for ASR in concrete: a new perspective, Eur. J. Environ. Civ. Eng. 23 (2019) 1311–1331. https://doi.org/10.1080/19648189.2017.1347068.

[61] M.T. De Grazia, N. Goshayeshi, R. Gorga, L.F.M. Sanchez, A.C. Santos, D.J. Souza, Comprehensive semi-empirical approach to describe alkali aggregate reaction (AAR) induced expansion in the laboratory, J. Build. Eng. 40 (2021). https://doi.org/10.1016/j.jobe.2021.102298.

[62] E. Lemarchand, L. Dormieux, F.J. Ulm, Micromechanics investigation of expansive reactions in chemoelastic concrete, Philos. Trans. R. Soc. A Math. Phys. Eng. Sci. 363 (2005) 2581–2602. https://doi.org/10.1098/rsta.2005.1588.

[63] B. Capra, J.P. Bournazel, Modeling of induced mechanical effects of alkali-aggregate reactions, Cem. Concr. Res. 28 (1998) 251–260. https://doi.org/10.1016/S0008-8846(97)00261-5.

[64] Y. Kawabata, K. Yamada, S. Ogawa, R. Martin, Y. Sagawa, S.E. Division, M. Division, S. Engineering, Correlation Between Laboratory Expansion and Field Expansion of Concrete : Prediction Based on Modified Concrete Expansion Test, 15th Int. Conf. Alkali-Aggregate React. (2016).

3.2   Introduction

Alkali-silica reaction (ASR) is a complex chemical reaction between un/poorly crystallized silica present in the aggregates and the alkali hydroxides (i.e., Na^+, K^+, OH^-) from the concrete pore solution [1–3]. ASR gel, a by-product of this chemical interaction, expands upon moisture uptake from the surrounding environment, resulting in the expansion and cracking of affected concrete [1,3–5]. Tensile stresses are generated withing the reactive aggregate particle as a consequence of

ASR-induced expansion [1]. As the reaction progresses, cracks which are initially generated within the aggregate particles tend to grow in number and size, and propagate to the cement paste, affecting the durability and serviceability of concrete infrastructure [6–10].

Over the last decades, severe cases of ASR in aging infrastructure (i.e., dams, bridges and others) took place worldwide, resulting in significant cracking, expansion and reduced capacity for resisting service loads, which raises important concerns for public safety [11,12]. In several cases, such scenarios have ended with the demolition of the affected structures as a safety precaution. Thus, appropriate protocols for the management of aging infrastructure are essential, helping to identify the mechanism at early stages, predicting the future development of deterioration, and allowing for the selection of better rehabilitation strategies. For ASR in particular, multiple guidelines have been published over the years, suggesting a variety of experimental testing procedures for evaluating the current condition (i.e., diagnosis) along with the potential for further deterioration (i.e., prognosis) [6,8,13,14] of affected concrete.

Amongst the proposed procedures and protocols, a comprehensive multi-level assessment tool based upon the coupling of microscopic and mechanical test procedures (i.e., particularly the Damage Rating Index - DRI and Stiffness Damage Test - SDT, respectively) was shown to be an efficient and reliable method to evaluate the damage degree of ASR-affected concrete [7,15]. This procedure has been implemented and deemed to work quite well on concrete specimens under "free-expansion" [7,15] and confinement conditions [16]; yet only a handful of field structures [6,17–19] have been appraised with such a protocol, showing promising results [17,18].

This work aims to use the multi-level protocol to appraise the condition of a selected number of deteriorated reinforced concrete columns from an ASR-affected concrete building after nearly 20 years of service. Furthermore, a comparison will be conducted between visual inspection and the multi-level assessment) to compare surface and internal damage due to ASR.

3.3   S.I.T.E. building

The School of Information Technology and Engineering (S.I.T.E.) building, located on the University of Ottawa campus (Ontario, Canada), was constructed in 2001 comprising of fifteen pyramid-shaped external rotunda columns providing structural support and stability to a central circular column located in the building's interior, as illustrated in Figure 3.2. A reinforced concrete pedestrian overpass leading to the topside entrance is supported by two of the columns, resulting in additional axial loads acting on the latter (Figure 3.2). The engineering drawings were the only technical documents recovered from the construction records; therefore, very few information regarding the concrete mix-design was available. According to the engineering drawings, concrete used for the columns was found to have a 28-day strength of 45MPa, with an air-entrainment of 5-7%.

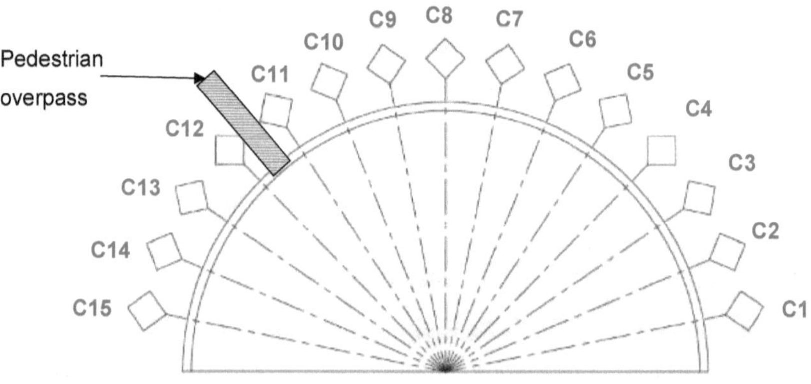

Figure 3.1: Labeled schematic of top view of columns.

Engineering drawings of the columns also reveal 3 layouts of reinforcement along the height of the column; Figure 3.3 illustrates the orientation of rebars and stirrups in a single column. The bottom portion comprises two layers of reinforcement: the outer layer consists of 12-35M rebars, whereas the inner layer incorporates 12-35M dowels (Figure 3.3C). The intermediate portion of the column is composed of 8-30M rebars (Figure 3.3B). Finally, the top portion of the column consists of 8-20M rebars (Figure 3.3A). 10M stirrups are spaced at 300mm in the bottom and intermediate portions, while a 75mm spacing is used at the top portion of the column.

Figure 3.2: A-D) Photos of the external rotunda columns of the S.I.T.E. building and E) top view sketch of columns.

A number of deterioration signs (e.g., map cracking and discoloration - Figure 3.2C; exudation of whitish products - Figure 3.2D, etc.) were observed at the various locations of distinct columns of SITE building, which raised concerns about the presence of internal swelling reaction mechanisms, particularly ASR. No important deflections were observed, although several cases of spalling were recorded. Moreover, surface scaling and delamination in some areas indicated that freeze-thaw (FT) damage could be present in the members.

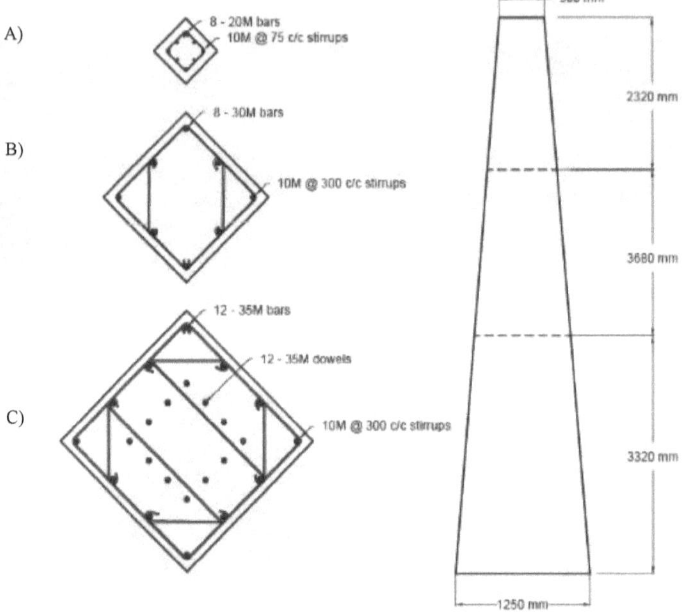

Figure 3.3: Reinforcement layout of external columns.

3.4   Diagnostic assessment tools

The aging of sound infrastructure is an inevitable process; thus, health monitoring protocols are essential to identify any problems that may negatively influence the structure's performance or safety. More so, the above protocols are of greater importance in the presence of distress mechanisms that compromise the material properties of sound concrete. Consequently, diagnosis, in the context of aging infrastructure, is the process of evaluating the cause and extent of concrete deterioration in order to quantify the inflicted "damage".

According to Sanchez et. al. [7], "damage" is defined as the harmful effects of distress mechanisms (e.g., ASR, freezing and thawing, etc.) on the durability, structural integrity and mechanical properties of affected concrete. Thus, damage in the context of this work refers to the stiffness reduction, loss in mechanical properties and reduction in the physical integrity of affected concrete members, thoroughly quantified through the coupling of mechanical testing and microscopic assessment.

Over the last decades, several microscopic and mechanical protocols, have been developed for the diagnosis of ASR-affected concrete. The cause of ASR damage is normally assessed by petrographic examination, which has the capacity to identify not only petrographic features of ASR (i.e., open cracks in aggregates and reaction product), but also the reactive aggregate responsible for distress [1]. The multi-level assessment protocol, based on microscopic (using the Damage Rating Index – DRI) and mechanical (using the Stiffness Damage Test – SDT) evaluation has also been successfully used to predict the extent of damage and reductions in mechanical properties (i.e., stiffness or strength) of ASR-affected concrete based on microscopic distress features (i.e., amount and pattern of internal cracking) [7,15,17,20].

3.4.1 Visual inspection

Visual inspection is the first step in the health monitoring of aging structures and is crucial to identifying potential deterioration mechanisms in concrete [21,22]. For a long time, visual inspection has been based on qualitative observations, thus evaluating the level of potential damage was a subjective process. However, in 1992, Godart et al. [23] proposed the cracking index (CI) as a protocol to evaluate the level of surface deterioration (i.e. extent of cracking) based on crack width and frequency (Eq. 3.1). The aim of the above protocol is identifying the need for more detailed investigations based on threshold values of 0.5 mm/m for the CI and 0.15 mm for average crack width, as per the ISE (1992) [13].

A)   B)

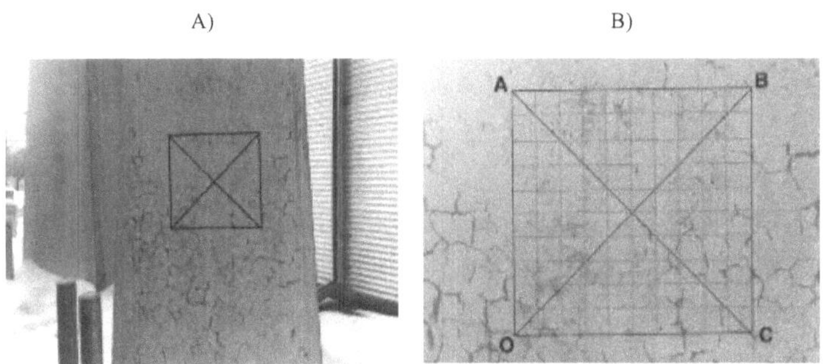

Figure 3.4: A) Cracking Index (CI) grid on column B) Closeup of CI grid.

$$CI = \frac{\sum crack\ opening}{Base\ length} \tag{3.1}$$

To calculate the CI, a 0.5x0.5 m grid is first constructed on the concrete surface of the most exposed (usually the most damaged) face (as per Figure 3.4A) and then each segment (i.e., vertical, horizontal and diagonal) is then divided into ten intervals (as per Figure 3.4B) [24–26]. The width of the largest crack intersecting each interval is then summed up to determine 'Σ crack opening'. Finally, the above value is divided by a 'base length' of 0.5 m to determine the CI.

3.4.2   Petrography

Petrographic examination is critical in the assessment of ASR damage since it possesses the diagnostic ability to confirm the presence of ASR-induced damage by identifying potentially reactive minerals and features associated with ASR damage (i.e., ASR-induced cracks and reaction products). This analysis is normally conducted through the use of a petrographic microscope under distinct optical conditions (i.e., using natural and polarized light) in accordance with ASTM C856 [27]. Supplementary tools including scanning electron microscopy (SEM), X-ray diffraction or infrared spectroscopy are often used to further examine the structure and composition of potentially reactive materials present in aggregates [1].

3.4.3   Multi-level assessment

3.4.3.1 Damage rating index (DRI)

Grattan-Bellew and coworkers [28] developed the Damage Rating Index (DRI), a semi-quantitative microscopic procedure used to assess damage in concrete. Damage features are counted in a 1x1cm grid drawn on the surface of concrete using a stereomicroscope under a 15-16X magnification. The distress features are then multiplied by the weighting factors which were proposed by Villeneuve & Fournier [29] in order to to balance their relative importance towards the corresponding distress mechanism (e.g., ASR). Figure 3.5 represents petrographic features and corresponding weighting factors consistent with DRI examination. The DRI is ideally carried out on a 200cm^2 polished concrete surface; however, for the comparison purpose the final result (i.e., DRI number) is normalized to 100cm^2. It is worth noting that a higher DRI number represents greater internal damage in the affected concrete [29].

A)   B)

Petrographic features		Weighing factors
Cracks in reactive aggregate	CCA	0.25
Opened cracks in reactive aggregate	OCA	2
Crack with reaction product in reactive aggregate	OCAG	2
Coarse aggregate de-bonded	CAD	3
Disaggregate/corroded aggregate particle	DAP	2
Crack in cement paste	CCP	3
Crack with reaction product in cement paste	CCPG	3

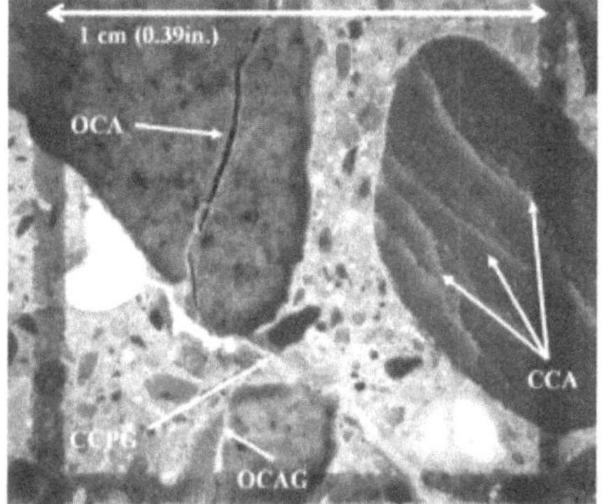

Figure 3.5: A) the microscopic distress features and corresponding weighting factors used in this study [29] and B) petrographic features associated with ASR damage [30].

The aim of proposed weighting factors is not only to distinguish the importance of various types of distress features but rather to reduce the variability between petrographers where for instance the same weighting factors is given for a crack with or without reaction product (i.e., ASR gel) since the recognition and interpretation might vary among distinct petrographers. Furthermore, analyzing the distinct weighting factors, one notices that the closed crack in the aggregate particles (CCA) is a distress features that might be related to any ASR-independent process such as aggregate crushing or weathering which a low weighting factor was given (i.e., 0.25) to those damage feature [31]. Increasing trend of open cracks with and without reaction products (i.e., OCAG and OCA, respectively) are clearly known as a function of ASR-induced damage [31]; hence, a weighting factor of 2 is attributed to such cracks. Finally, due to the significance of cement paste in governing the mechanical properties of concrete, cracks in cement paste with and without reaction product (i.e., CCPG and CCP, respectively) pertain to a weighing factor of 3.

3.4.3.2 Mechanical tools

The Stiffness Damage Test (SDT) is a mechanical tool used to appraise condition of concrete affected by internal swelling reaction (ISR) mechanisms (e.g., ASR, FT or DEF) [32–34]. This testing procedure was initially developed by Walsh, who noticed a good correlation between the crack density and the cycles of loading/unloading of rock specimens [35], before Crouch adapted the testing procedure for concrete specimens [36]. An initial fixed loading of 5.5MPa (Crisp et al. [37,38]) was used to test concrete specimens over five cycles of compressive loading. After a thorough testing on a wide range of reactive aggregates, Smaoui et al. [39] proposed adjusting the fixed loading to 10MPa for the quantification of ASR-induced damage. Finally, Sanchez et al. [33,34] proposed that the SDT be should be carried out using a percentage (40%) of the compressive strength (ultimate capacity) of the concrete specimens instead of fixed load to improve the diagnostic character of the test; the Stiffness Damage Index (SDI) and Plasticity Deformation Index (PDI), also proposed by the same authors [33,34], represent the ratio of dissipated energy/plastic deformation to the total energy/deformation of concrete specimens (i.e., SI / SI+SII and DI / DI+DII) over the five cycles (Figure 3.6). Furthermore, the modulus of elasticity (ME - average secant modulus of the 2nd and 3rd cycles) as well as the non-linearity index (NLI - Sec 1 divided by Sec 2 in first cycle – as per Figure 3.6) as proposed by Crisp et al. [37,38] could also be deemed efficient SDT outcomes for detecting the damage extent and orientation (i.e., NLI is higher/lower than one for cracks oriented perpendicular/parallel to loading), respectively, in affected concrete.

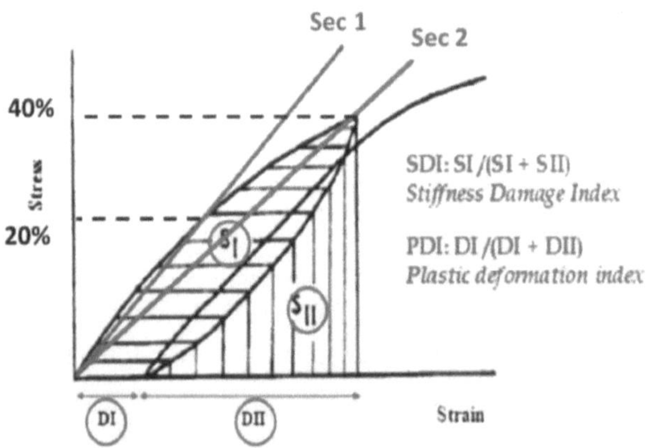

Figure 3.6: Calculation of the SDI and PDI as per [40].

3.5   Scope of work

Important deterioration signs were visually observed on the external surfaces of a number of S.I.T.E. building columns, which raised concerns about the presence of internal swelling reactions, particularly ASR. Therefore, a comprehensive condition assessment protocol, comprising visual inspection, petrography and the multi-level assessment as per [17,41] was conducted with the aim of diagnosing the cause and extent of damage affecting the building. The multi-level assessment consists of series of mechanical testing and microscopic assessment, through the use of the Stiffness Damage Test (SDT) and Damage Rating Index (DRI), respectively, to quantify ASR-induced damage. The above tools were used to evaluate a number of specimens extracted from four columns, selected based on the comprehensive visual inspection conducted to identify columns displaying advanced signs of surface deterioration. The extent of damage and implications on material properties of the selected columns, as well as the influence of exposure conditions and restraining factors on external and internal concrete deterioration will be discussed.

3.6   Materials and methods

3.6.1   Visual inspection

The visual inspection was conducted on all fifteen columns of the SITE building by determining the corresponding CI. Once the most exposed face was identified, the CI grid was constructed on the concrete surface as per Figure 3.4. To maintain the integrity when comparing the columns, the CI grid was constructed in a central position at a height of 1.50 m from the ground for each column. Finally, the CI was calculated as per Eq. 2.1.

3.6.2   Coring

For each of the columns selected through visual inspection, two longitudinal cores (at different heights) were drilled along the width of the column. Based on the recommendation of the structural report (as per structural analysis in Appendix A and Appendix B) prepared to ensure the safety of the procedure, an interval of one meter was allowed between the two cores (Figure 3.7). The surface was then scanned using Ground Penetrating Radar (GDR) concrete scanner to identify the depths and locations of rebars, so that the safest locations for coring may be identified (i.e., as per Figure 3.7). Afterwards, samples were drilled through the use of 100mm-diameter hydraulic drills stabilized using industrial clamps against the surface in order to extract perfectly horizontal specimens as shown in Figure 3.8. Afterwards, the cores were cut into 200mm-length specimens using a wet concrete saw, promptly wrapped in plastic film and then stored in an environmentally controlled chamber at 12°C ± 2°C to stop further deterioration as per [7] (due to limited testing capacity of laboratory setups).

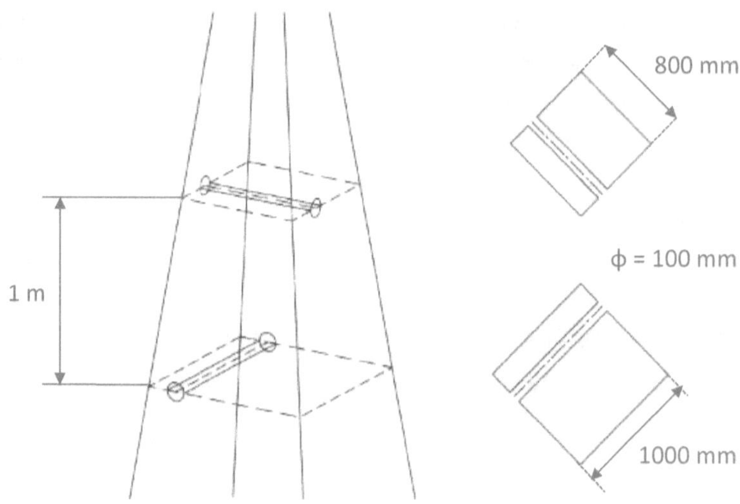

Figure 3.7: Depiction of estimated coring locations in columns.

Figure 3.8: Equipment installation and subsequent core drilling.

3.6.3 Laboratory test methods

3.6.3.1 Petrographic examination

Petrographic examination was carried out on the specimens extracted from columns presenting extensive visual deterioration signs with two different, yet related goals. First, to identify the pathologies that may be present by examining petrographic features including microcracks in aggregate particles, cement paste or cement paste-aggregate bonds, and second, to determine the mineralogy of the concrete mix (i.e., rock types/mineral phases of aggregates, cement type, volumetric proportions, etc.); this step is important in identifying the source of existing pathologies. The analysis was conducted on thin sections (75x50mm) through optical microscopy (with natural and polarized light) along with supplementary SEM analysis in accordance with ASTM standards [42].

3.6.3.2 Damage rating index (DRI)

The DRI was conducted on three samples per column; thus, the test results for each column represent the average of three samples. Prior to examination, the samples were cut in half using a diamond bladed masonry wet saw and then mechanically polished using grits 30 (coarse), 60, 140, 280, 600, 1200, and 3000 (very fine) to achieve a clear and reflective surface. Once sufficiently polished, a 1 by 1cm grid was manually constructed on the surface before proceeding with the microscopic examination. Afterwards, petrographic features were examined under a 15-16X magnification in accordance with [31,43], and cracks were multiplied by the relevant weighting factors previously displayed in Figure 3.5.

3.6.3.3 Mechanical tools

Compressive strength (CS) testing was carried out on two internal core specimens extracted from the least damaged column as per the visual inspection. The SDT, on the other hand, was conducted on three samples per column; thus, test results represent the average of three samples. The aim of the CS test was to identify any reduction in concrete strength compared to the 28-day design strength of the original mix as well as determining the 40% strength value necessary to conduct the SDT. Core specimens were grinded in preparation for both test procedures in accordance with CSA23.2-9C [44] to even out the contact surface to ensure a homogeneous stress distribution and therefore accurate test results. SDT samples were also conditioned in an environment of 23°C ± 2°C and 100% relative humidity (RH) for a period of 48h in accordance with CSA 23.2-14C [45] and as per Sanchez et al. [34], thus enabling sample rewetting and reducing the test's variability. Afterwards, the SDT was conducted following the method proposed by Sanchez et al. [7,33]: five cycles of loading up to 40% of the ultimate strength of sound concrete at a constant loading rate of 0.10 MPa/s.

3.7   Results

3.7.1   Visual inspection

The most notable sign of surface damage observed in all columns was map cracking surrounded by 'dark rims' of moisture (Figure 3.9A-C), which is a signature of ASR damage. Moreover, exudations of whitish products were discovered in some columns (as seen in Figure 3.9D and E). Other signs of deterioration including spalling (Figure 3.9F) and surface scaling were also observed.

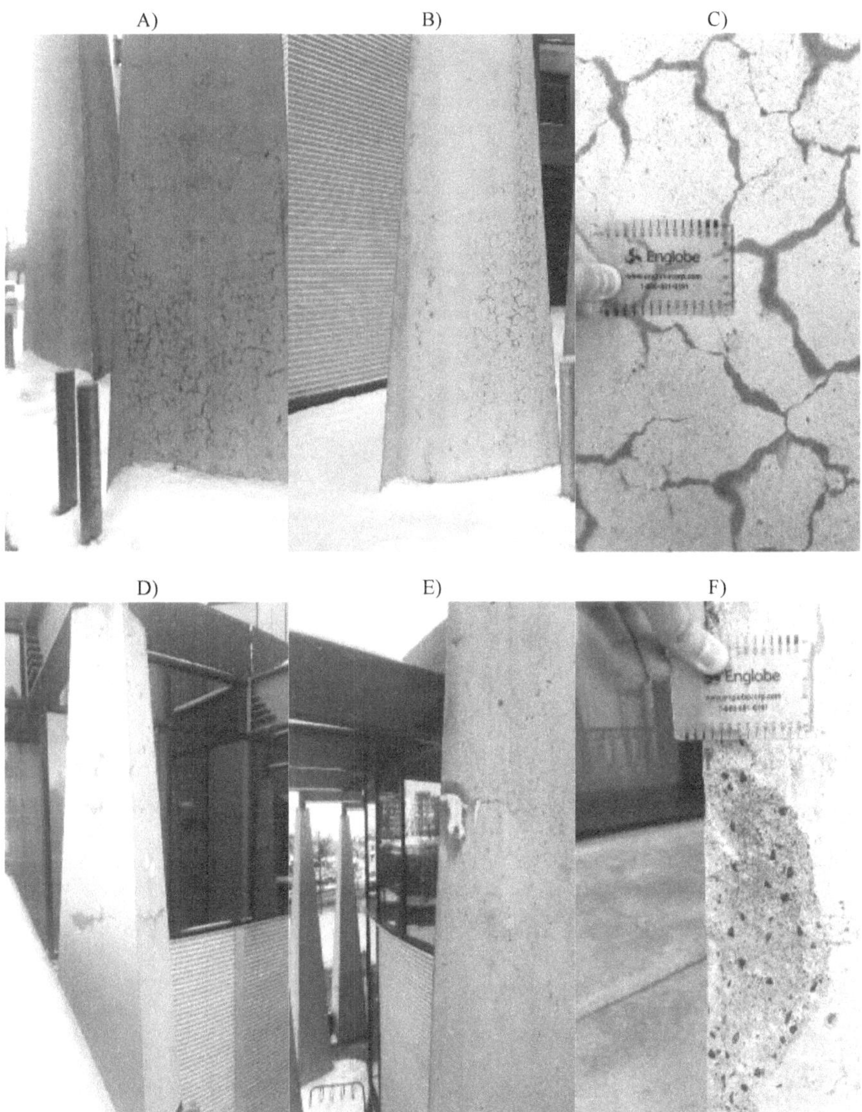

Figure 3.9. Surface signs of ASR: (A – C) Numerous instances of map cracking (D – E) exudations of white product F) spalling.

Figure 3.10A and B show the CI and crack measurements recorded for all columns, respectively. Results demonstrate that further laboratory investigation for ASR is required for columns C1, C2, C3, C4, C7, and C11 since the above columns were found to have a CI > 0.5 mm/m and average crack width > 0.15 mm. However, the four columns with the greatest CI and crack width (i.e., C1, C3, C4, and C7) were selected for core drilling. Finally, based on the observed signs of severe surface deterioration (e.g., Figure 3.9), C1 and C4 were assumed to be 'very highly' damaged, whereas C3 and C7 were assumed to be 'highly' damaged.

A)

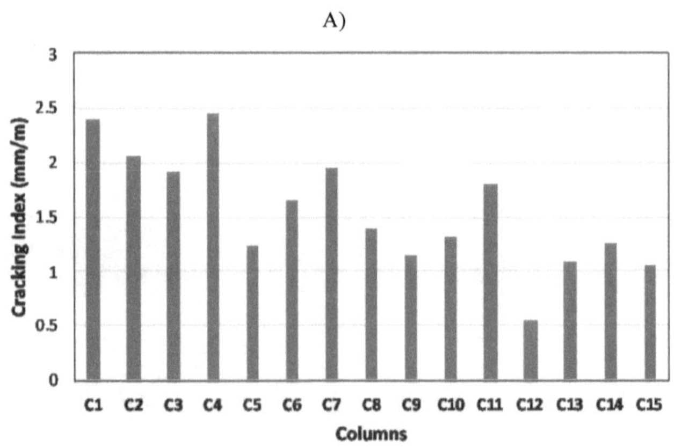

B)

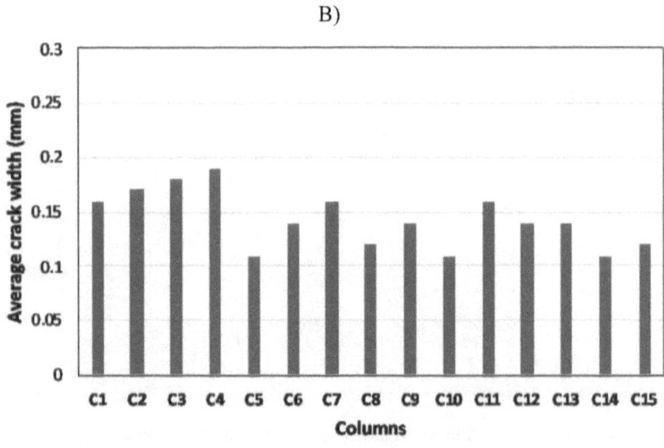

Figure 3.10: Visual inspection of columns: A) Cracking Index (mm/m) B) average crack width (mm).

3.7.2   Petrographic examination

Petrographic examination was carried out on thin sections prepared using the cores extracted from C1 to evaluate the cause(s) of deterioration affecting S.I.T.E. building. C1 was selected since it was presumed to be the most damaged column among the selected members based on visual inspection results (i.e., highest sign of damage through convention visual inspection as well as a high CI value - 2.40 as per Figure 3.10A). Moreover, since the deterioration signs verified on the various columns were not different from one another (only the degree has varied), the evaluation of cores extracted from C1 was considered sufficient to understand the cause(s) of deterioration of S.I.T.E building, particularly on its columns. Upon the examination of the thin sections, various cracks were verified both in the coarse aggregate particles and cement paste (Figure 3.11), mostly originating from coarse aggregates. Furthermore, the composition of coarse aggregate particles revealed to presence of dolomitic limestone and mudstone, containing minor amounts of microcrystalline quartz. ASR gel was also observed in the cracks in the coarse aggregate as well as cement paste (Figure 3.11A and B); similarly, reaction products were observed in cement paste voids (Figure 3.11C and D).

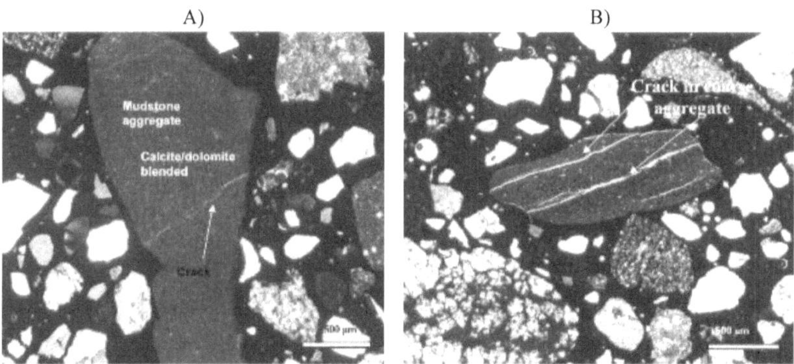

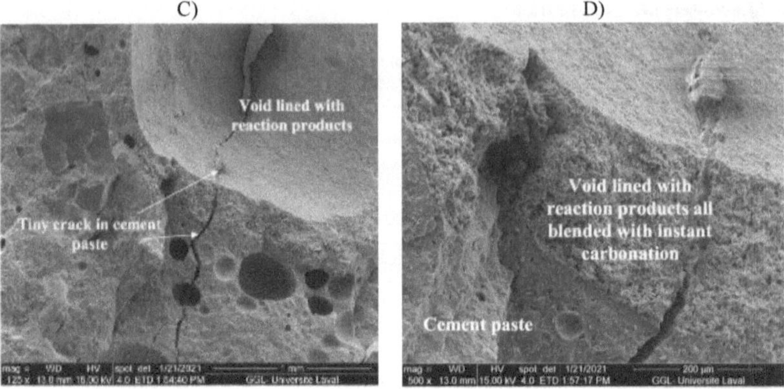

Figure 3.11: A) Coarse aggregate (mudstone) under 25X magnification (polarized light), B) Crack in coarse aggregate under 25X magnification (natural light), C) Crack in cement paste and void under SEM (secondary electron analysis) and D) Blended reaction product with calcite crystals on surface under SEM (secondary electron analysis).

3.7.3   Damage rating index (DRI)

The results of DRI analysis performed on the core specimens retrieved from distinct concrete columns are displayed in Figure 3.12. Evaluating the plot, one observes that the most important damage features noticed in this study are the closed cracks in the aggregates (CCA - blue chart), the open cracks in the aggregates (OCA - red chart) and the cracks in the cement paste (CCP - orange chart). All specimens were found to have a quite similar number of closed cracks in the aggregate particles (i.e., C1=120, C3=129, C4=100, C7=110); it is worth noting that the closed cracks in the aggregates (CCA) are not necessarily related to deterioration mechanisms such as ASR, but rather to crushing or weathering processes of aggregates. Otherwise, the open cracks in the aggregates (OCA) and cracks in the cement paste (CCP) varied significantly as per the column appraised. The highest CCA number was found in C4 (i.e., 132), followed by C3 (i.e., 76), C7 (i.e., 46) and C1 (i.e., 46). Likewise, the greatest CCP number was also observed in C4 (i.e., 88) and C3 (i.e., 45), while C1 and C7 displayed similar CCP values (i.e., 21 and 20, respectively). Furthermore, cracks containing reaction products were limited, yet still visible during the microscopic examination. As such, C4 was verified to contain the greatest number (i.e., 51) of open cracks in the aggregates with gel/reaction product (OCAG - green chart in Figure 3.12), followed by C3 (i.e., 14); CCPs were observed in C1 and C7. Finally, few cracks (i.e., 9) were also observed in the cement paste containing gel/reaction products (CCPG - light blue chart in Figure

45

3.12), particularly in C4. Based upon the above observations, the highest DRI number was found in C4 (i.e., 381) followed by C3 (i.e., 265), while C1 and C7 yielded similar DRI values (i.e., 187 and 176, respectively).

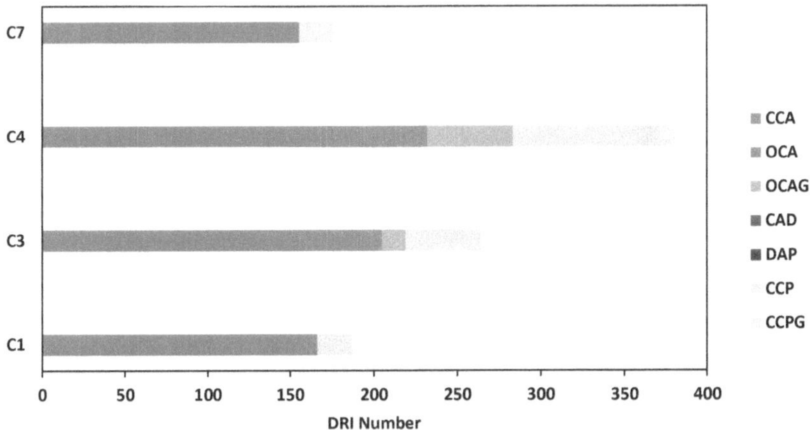

Figure 3.12: Damage Rating Index (DRI) results from the four examined columns.

3.7.4 Stiffness Damage Test (SDT)

The CS test was conducted using two specimens retrieved from C7 as it appeared to be the least damaged column out of the four selected as per the visual inspection (i.e., lowest visual signs of deterioration and CI value - 1.95 as per Figure 3.10A). Concrete strength was therefore determined to be 55 MPa which showed a higher value compared to the columns design strength (i.e.,45 MPa). Figure 3.13 illustrates the SDI, PDI and ME values gathered from the core specimens from C1, C3, C4 and C7. As previously stated, results from each column represent the average values (i.e., SDI, PDI and ME) determined from three core specimens per column.

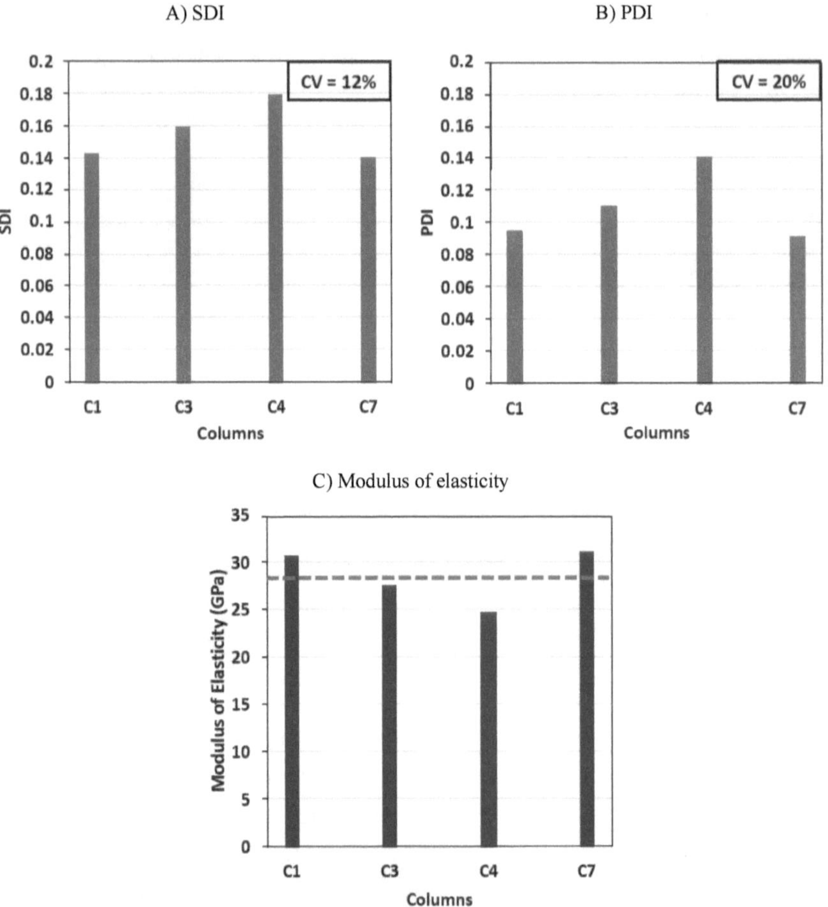

A) SDI   B) PDI

C) Modulus of elasticity

Figure 3.13: Stiffness Damage Test (SDT) results: A) SDI, B) PDI and C) Modulus of elasticity (GPa).

It is important to note that higher levels of distress are linked with higher SDI and PDI, as well as lower ME [7]. According to Figure 3.13A, one observes that the highest SDI was observed for C4 (i.e., 0.18) followed by C3 (i.e., 0.16), whereas the smallest SDI value was reported for C1 and C7 (i.e., 0.14 for both). Similarly, the highest PDI was gathered on C4 (i.e., 0.14) followed by C3 (i.e., 0.11), whereas only a negligible difference was noticed between C1 and C7 (i.e., 0.10 and 0.09, respectively). Moreover, comparing the ME of distinct columns, one notices that C1 and C7 exhibits the highest values for ME (i.e., 31 GPa) followed by C3 (i.e., 28 GPa) and then the lowest ME was reported for C4 (i.e., 25 GPa). The average ME was determined to be around 28 GPa, as shown in Figure 3.13C (i.e., red line)

3.8   Discussion

3.8.1   Damage degree and induced expansion

The condition assessment was conducted on core specimens extracted from the external columns of the SITE building (i.e., Ontario, Canada) to appraise the cause(s) and extent of damage in the affected members. According to the petrographic examination performed on the concrete specimens extracted from C1, various distress features associated with ASR including cracks in coarse aggregates (Figure 3.11A and B), and the presence of reaction product (i.e., ASR gel) in the coarse aggregate particles and cement paste (Figure 3.11B and D, respectively) clearly confirmed the presence of ASR in the SITE building. Nevertheless, due to the presence of somewhat minor signs of ASR in the thin sections, the chemical reaction was deemed to be in the early stages and thus the induced deterioration of the affected members was believed to be low.

Similar to the results obtained through the petrographic analysis, the results of microscopic assessment (Figure 3.12) revealed signs of ASR-induced damage in specimens collected from all examined columns (i.e., C1, C3, C4, and C7). As such, the presence of open cracks without and with reaction product (i.e., OCA and OCAG - red and green charts, respectively) as well as a small number of cracks in the cement paste containing ASR gel (i.e., CCPG - light blue chart) could be a clear indication of ASR in all concrete columns of this study. Accordingly, Figure 3.14 illustrates the above features in examined specimens, including the presence of ASR gel in aggregate particle and cement paste cracks. It is worth mentioning that the number of cracks observed in the cement paste are higher than one may expect for an ASR reactive coarse aggregate as observed by Sanchez et al. [15]. In the absence of reactive agents, cracking in cement paste may be a sign of additional distress mechanisms (e.g., internal or external sulphate attack, freeze and thaw, shrinkage, etc.). However, given the location (i.e., Canada) and environmental conditions to which the columns are exposed, one may consider that freeze and thaw (FT) damage is likely present in the affected elements [46–48]. Therefore, deterioration in the assessed columns is likely due to the effect of a combined distress mechanism (i.e., ASR combined with FT) as previously reported by Sanchez et al. [15].

A) Open crack with gel in reactive coarse aggregate

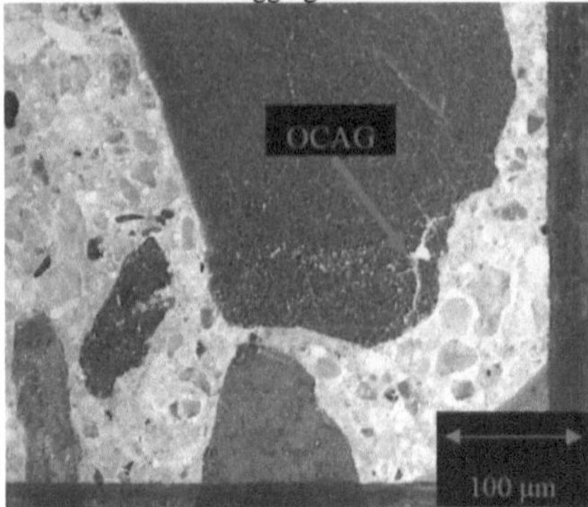

B) Open crack with gel and closed crack

C) Cement paste cracks with gel extending from reactive aggregate

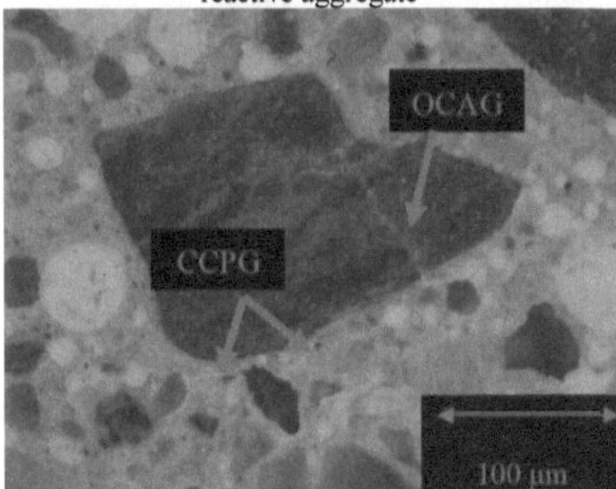

D) Crack in the cement paste with gel

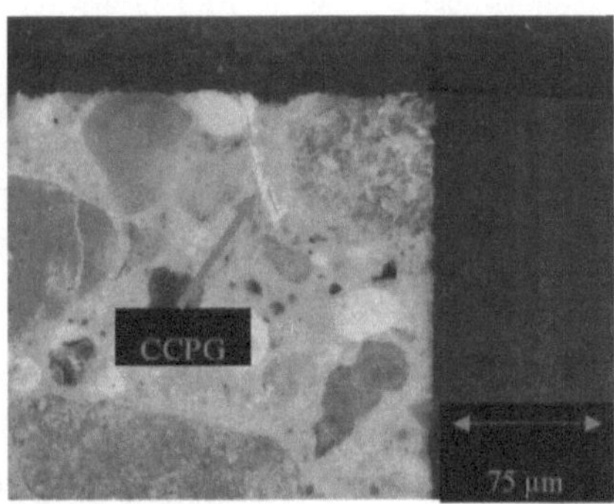

Figure 3.14: Petrographic features associated with ASR expansion observed through DRI analysis.

Having discussed the microstructure of the examined specimens, one may interpret the effect of ASR damage on their mechanical behaviour. All specimens from this study (i.e., C1, C3, C4, and C7) demonstrated various levels of mechanical damage, with C4 displaying the greatest SDI, PDI, and lowest ME (Figure 3.13). Furthermore, the compressive strength observed for C7 specimens was found to be greater than the design strength (i.e., as shown in section 3.7.3), and thus not yet a matter of concern from a serviceability viewpoint. This was in accordance with findings published by the Institution of Structural Engineers (ISE, 1992) [13], which mentioned that in most ASR-affected structures, the compressive strength loss is normally smaller than the gap between the design strength and "actual" strength; the above is especially true for low to high expansion levels (i.e., 0.05 – 0.20% expansion) [7,49]. On the other hand, the modulus of elasticity tends to be more drastically affected at earlier stages of the reaction (i.e., ≈ 0.05% expansion) [7,13,15,49,50]. Despite the lack of data from undamaged concrete, notable losses in the elastic modulus are observed when the columns exhibiting higher damage (i.e., higher SDI and PDI from C3 and C4) are compared to those presenting lower distress (i.e., lower SDI and PDI from C1 and C7); C3 and C4 displayed approximately 14% and 22% lower ME, respectively.

In previous studies conducted by Sanchez et al. [7,15], a multi-level assessment protocol was developed by the authors, integrating microscopic and mechanical (i.e., DRI and SDT) analyses to evaluate the degree of damage of ASR-affected concrete. This protocol was developed with 95% confidence using 20 mixtures incorporating a wide range of concrete strengths (i.e., 25MPa, 35MPa, 45MPa) and 13 distinct reactive aggregates (i.e., coarse and fine) to classify the damage degree (mechanical and microscopic) of ASR-affected concrete as a function of its induced expansion [7,15]. A four-quadrant chart (i.e., plot of expansion, SDI, loss in ME – δ, and DRI number) was developed for each mixture based upon results from laboratory testing (i.e., SDT and DRI). Table 3.1, developed as part of the aforementioned study, summarizes and classifies the damage degrees (i.e., expansion levels) according to the charts developed for all mixtures of concrete specimens under "free expansion" conditions. A good correlation has been observed between induced expansion, microscopic (i.e., DRI number), and mechanical (i.e., SDI and δ) results, regardless of the type and nature of the reactive aggregate and concrete strength. Despite being developed under "free" expansion conditions [7,15], studies using the multi-level assessment for confined laboratory concrete [16] and in-situ core specimens [6,17–19] showed promising results. Moreover, diagnosis through the multi-level assessment may be used as a reference point for field concrete, and even assumed to be the "worst case scenario". Thus, the proposed protocol is deemed efficient to appraise the condition of ASR-affected concrete structures in the field (subject to various confinement and environmental conditions).

Table 3.1: Multi-level assessment results of AAR affected concrete [7].

Classification of ASR damage degree (%)	Reference expansion level (%)	Assessment of ASR				
		Stiffness loss (%)	Compressive strength loss (%)	Tensile strength loss (%)	SDI	DRI
Negligible	0.00 – 0.03	-	-	-	0.06 – 0.16	100 – 155
Marginal	0.04 ± 0.01	5 – 37	(-) 10 – 15	15 – 60	0.11 – 0.25	210 – 400
Moderate	0.11 ± 0.01	20 – 50	0 – 20	40 – 65	0.15 – 0.31	330 – 500
High	0.20 ± 0.01	35 – 60	13 – 25	45 – 80	0.19 – 0.32	500 – 765
Very high	0.30 ± 0.01	40 – 67	20 – 25		0.22 – 0.36	600 – 935

As per Sanchez et al. [7,15], practitioners may deploy the charts for the diagnosis of ASR-affected concrete in one of two ways: 1) selecting the envelope of the same/ similar aggregate to the one being examined or 2) averaging the envelopes in the event that the reactive aggregate is unknown. Since the reactive aggregate was observed to be a limestone type aggregate through petrographic examination, the multi-level assessment charts will be used in accordance with the first proposed method for this study. Figure 3.15 represents the global assessment chart for concrete mixtures of distinct strengths (i.e., 25, 35 and 45 MPa), incorporating a coarse reactive limestone aggregate and non-reactive sand. Based on the envelope of the 45 MPa concrete (same as design strength of columns) displayed in Figure 3.15 and Table 3.1as per Sanchez et al. [7,15], the greatest induced expansion was obtained for C4 (i.e., moderate – 0.08%), followed by C3 (i.e., marginal – 0.05%) and C1/C7 (i.e., negligible – 0.03%) as displayed in Figure 3.16.

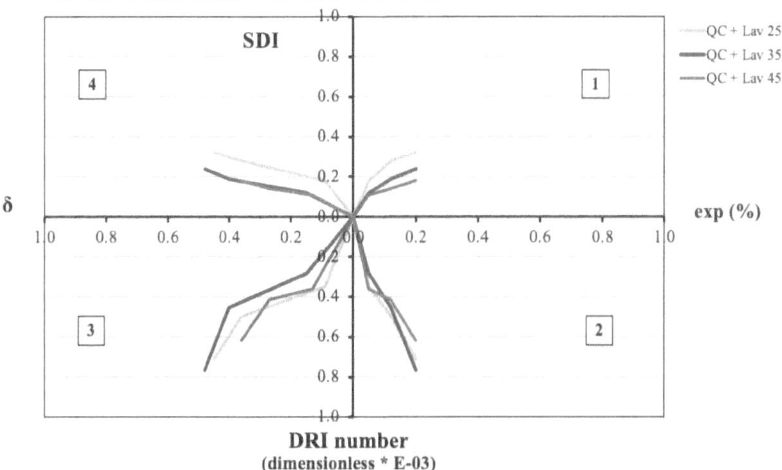

Figure 3.15: Global assessment of concrete containing similar reactive aggregate (limestone type) to the coarse aggregate in the S.I.T.E. column mix [15,17].

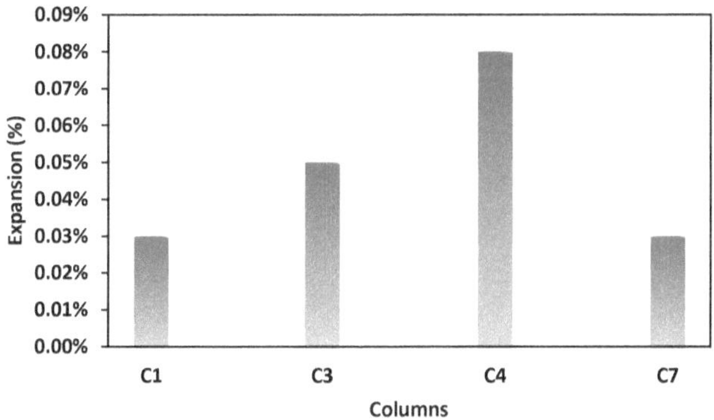

Figure 3.16: Potential expansion attained for assessed columns.

One of the primary goals for condition assessment is determining the impact of distress on the serviceability of the affected structure; this could be done using charts similar to the one presented in Figure 3.15 that are able to predict, besides the expansion attained to date, mechanical properties losses of affected concrete as a function of ASR damage. According to the expansion levels presented in Figure 3.16, the conditions of appraised members (i.e., C1, C3, C4, and C7) do not raise immediate concerns in terms of structural performance or serviceability (i.e., deformation, major cracking, spalling, etc.). However, serviceability problems are anticipated in the short to middle terms (i.e., within 5 to 15 years), particularly in C3 (i.e., 0.05%) and C4 (i.e., 0.08%), since the multi-level chart (brown chart - Figure 3.15) indicates a sharp drop (i.e., $\geq$ 15%) in ME past an expansion of 0.05%. Thus, immediate rehabilitation measures (e.g., surface treatment) and periodic monitoring are recommended for all columns [6,8,14]. Finally, it is important to notice that ASR-induced development presents an anisotropic behaviour under confinement due to the so-called "expansion transfer" phenomenon, since expansion reduction in the confined direction is often counteracted by an increase in expansion along other directions [16,17,39,51–56]. Therefore, since the evaluated core specimens were extracted from less confined locations, one needs to be aware of the likely influence of this phenomenon (i.e., higher deterioration values when compared to more confined locations) on the multi-level results gathered in this research.

3.8.2 Visual versus multi-level assessment

Typically, visual inspection is the first step for health monitoring of aging concrete infrastructure [21,22]. As previously mentioned, surface crack mapping techniques, such as the CI, allows one to exploit crack width and frequency to evaluate surface damage of affected concrete [6,13,23]. Despite a lack of literature demonstrating the capacity of the CI to evaluate the degree of damage of ASR-affected concrete, it serves as a reliable tool for justifying further analysis through microscopic and/or mechanical testing [41].

Comparing the results of the visual inspection performed through the CI, and the actual damage degrees of columns C1, C3, C4 and C7, determined through the multi-level assessment protocol (i.e., Figure 3.16), one may infer that there is no correlation between the two sets of data, as shown in Table 3.2. These results are consistent with the observations made from the condition assessment of an overpass affected by ASR previously [17], where distinct visual inspection methods failed to accurately determine the "inner extent" of damage. Based on comparisons from Table 3.2, evaluation of damage through visual inspection seems to have two inconsistencies: 1) overestimation of damage, and 2) inability to distinguish different classes of induced deterioration. For instance, visual inspection findings indicate that C1 is more damaged than C3 (i.e., CI of 2.40

and 1.92, respectively as per Figure 3.10), yet the opposite was verified as per the multi-level assessment (i.e., 0.03% and 0.05% expansion, respectively as per Figure 3.16). Likewise, a number of researchers [24–26,57,58] have observed conflicting results when implementing the CI; the latter could underestimate ASR expansion in laboratory settings (i.e., in some cases as much as 40%) [26,57,58]. Overall, the mismatch between the two sets of data attest that the data obtained through the visual inspection may not represent the real internal concrete damage of the affected member. Although limited, yet interesting results on the diagnostic potential of CI for detecting the presence of ASR in affected concrete are displayed in the literature, to the authors' best knowledge, there is still a lack of research to systematically correlate the crack importance (i.e., width and length) obtained through visual inspection and the internal damage degree of affected concrete. Thus, further work in the above area could produce useful findings for the damage assessment of aging concrete through crack-mapping techniques.

Table 3.2: Comparison between visual inspection and the multi-level assessment.

Column	Visual Inspection			Multi-level assessment		
	CI (mm/m)	Average crack width (mm)	Damage Degree	Expansion (%)	Damage Degree	Correlation ($\checkmark$/X)
C1	2.40	0.16	Very high	0.03	Minimal	X
C3	1.92	0.18	High	0.05	Marginal	X
C4	2.45	0.19	Very high	0.08	Moderate	X
C7	1.95	0.16	High	0.03	Minimal	X

Although a poor correlation between visual inspection and CI results was observed in this study, interesting trends were observed for the recorded crack widths. On one hand, evaluating the crack importance (i.e., width) of distinct concrete elements of this study, one notices that C4 and C3 demonstrated the greatest values (i.e., 0.19 and 0.18, respectively) followed by C1 and C7 (i.e., 0.16 for both). On the other hand, as per the multi-level assessment (Figure 3.16), the greatest degree of damage was observed in C4 (i.e., moderate - 0.08%) followed by C3 (i.e., marginal - 0.05%) and C1/C7 (i.e., negligible for both - 0.03%). Thus, the above observation suggests that larger crack widths may be associated with greater internal damage due to ASR.

3.8.3 Coupled damage mechanisms

An increasing number of cracks in the cement paste was observed in some specimens (i.e., C3 and C4) during microscopic examination. As previously mentioned in Section 3.8.1, along with open cracks in reactive aggregates, such petrographic features signify a potential combined distress mechanism, featuring ASR coupled with freezing and thawing (FT). One of the ways to confirm the presence of FT is to identify the location of origin of cracks observed in the cement paste. Normally, FT-related cracks are generated and propagate in the cement paste [46,48,59–65]. Figure 3.17 displays the number of cement paste cracks observed originating from either reactive aggregates or bulk cement paste for each of the examined specimens. Moreover, Figure 3.18 shows the image of a crack originating in the cement paste.

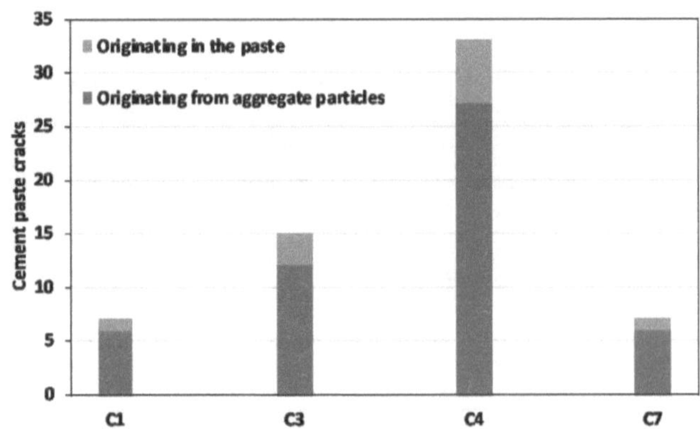

Figure 3.17: Origin of cement paste cracks.

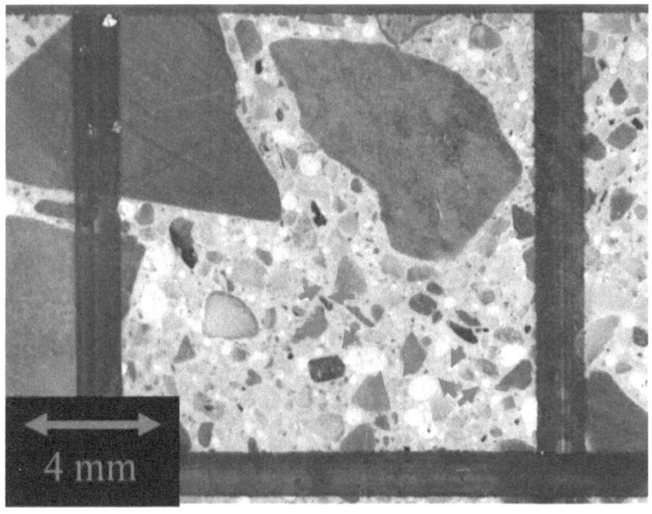

Figure 3.18: Crack originating in the cement paste.

The microscopic analysis of concrete affected by uncoupled and coupled FT (i.e., ASR+FT) conducted in previous works revealed greater DRI numbers than samples affected only by ASR [7,15,17]. This is particularly evident in the condition assessment of of the Robert-Bourassa/Charest (RBC) bridge, where a group of columns demonstrated significantly more cracks in the cement paste and less cracks in reactive aggregates than identical columns supporting a different area of the bridge, despite comparatively similar DRI numbers in all columns [17]. By further analysing specimens from S.I.T.E. columns, one could infer that the percentage of crack originating in the cement paste as a function of the total cracks in the cement paste is highest in C3 (i.e., 25%), followed by C4 (i.e., 23%) and C1/C7 (i.e., 17%), which shows that FT is likely present in C3 and C4. Furthermore, considering the multi-level assessment of the columns, it appears that these percentages are higher in more damaged columns (i.e., C3 – 0.05%, C4 – 0.08%), which could indicate that FT distress could continue to increase in the columns as a function of increasing ASR damage. In fact, Bérubé & Deschenes [48,66,67] concluded that distress features (i.e., cracking) generated by ASR facilitate the moisture ingress into cement paste voids, thus accelerating the rate of deterioration caused by FT mechanism. Thus, it is evident that

FT is developing in the columns at greater expansion levels, and one could expect a significant increase of cement paste cracks as expansion increases.

Research has shown that significant reductions in mechanical properties (e.g., up to 40%, 73% and 52% reduction of compressive strength, tensile strength and modulus of elasticity as per [15]) of damaged concrete could occur with FT and coupled ASR+FT [15,46,63,64]. Contrary to ASR, which raises concerns regarding structural serviceability (i.e., serviceability limit state – SLS, related to deflections and stiffness loss), FT raises concerns regarding structural safety (i.e., ultimate limit state – ULS) due to the nature of damage which occurs primarily in the cement paste and ultimately reduces concrete strength [15]. Therefore, significant structural concerns could develop in C3 and C4 within the next 10 years. However, such problems will likely take longer to develop in C1 and C7, given their lower level of expansion.

3.9   Conclusion

The purpose of this study was to assess the condition of the deteriorated external columns of a building in University of Ottawa (S.I.T.E. building located in Ontario, Canada) after around 20 years in service. Initially the visual integrity (i.e., surface damage) of distinct concrete columns was evaluated through the various visual inspection methods and later four columns displaying the highest sign of deterioration were selected for future analysis. Finally, petrographic analysis (i.e., thin section-SEM) and multi-level assessment (i.e., mechanical and microscopic examination) was performed on extracted concrete specimens to identify the cause and extent of damage in distressed concrete elements. The following conclusions can be drawn from the findings of this work:

- The petrographic examination, through the SEM, confirmed the presence of ASR in the analyzed concrete, as ASR gel was found in cement paste voids and lining some cracks emerging from the coarse aggregate particles. Further examination under natural and polarized light concluded that the coarse aggregate was comprised primarily of variations of alkali-silica reactive dolomitic limestone and mudstone.

- A comprehensive microscopic (i.e., DRI) and mechanical (i.e., SDT) testing protocol through the use of multi-level assessment demonstrated that despite close proximity and essentially similar exposure conditions, different levels of damage were found in the columns (i.e., 0.03% for C1 and C7, 0.05% for C3, and 0.08% for C4) tested in this study.

- Similar to the previous works, no correlation was found between visual inspection classification and internal damage presented in the examined columns. In addition to overestimating the degree of internal damage due to ASR, the CI was also unable to distinguish between different levels of damage. Nevertheless, average crack widths showed similar trends to the damage degree attained using the multi-level assessment.

- A more detailed analysis of microscopic observations revealed that FT could be developing in some columns (i.e., C3 and C4) at higher levels of expansion, raising important concerns in the short-term. Further investigation into the potential source of FT could provide insight into potential mitigation measures. Regardless, rehabilitation measures and periodic monitoring of the columns is required based on the current diagnosis to properly manage the distress and prevent further expansion.

3.10 References

[1]   B. Fournier, M.A. Bérubé, Alkali-aggregate reaction in concrete: A review of basic concepts and engineering implications, Can. J. Civ. Eng. 27 (2000) 167–191. https://doi.org/10.1139/l99-072.

[2]   F. Rajabipour, E. Giannini, C. Dunant, J.H. Ideker, M.D.A. Thomas, Alkali-silica reaction: Current understanding of the reaction mechanisms and the knowledge gaps, Cem. Concr. Res. 76 (2015) 130–146. https://doi.org/10.1016/j.cemconres.2015.05.024.

[3]   S. Chatterji, Chemistry of alkali-silica reaction and testing of aggregates, Cem. Concr. Compos. 27 (2005) 788–795. https://doi.org/10.1016/j.cemconcomp.2005.03.005.

[4]   S. Chatterji, N. Thaulow, A.D. Jensen, Studies of alkali-silica reaction. part 5. Verification of a newly proposed reaction mechanism, Cem. Concr. Res. 19 (1989) 177–183. https://doi.org/10.1016/0008-8846(89)90081-1.

[5]   S. Chatterji, N. Thaulow, A.D. Jensen, Studies of alkali-silica reaction, part 6. Practical implications of a proposed reaction mechanism, Cem. Concr. Res. 18 (1988) 363–366. https://doi.org/10.1016/0008-8846(88)90070-1.

[6]   B. Fournier, M.A. Berube, K.J. Folliard, M. Thomas, Report on the Diagnosis, Prognosis, and Mitigation of Alkali-Silica Rection (ASR) in Transportation Structures, Austin, 2010.

[7]   L.F.M. Sanchez, B. Fournier, M. Jolin, D. Mitchell, J. Bastien, Overall assessment of Alkali-Aggregate Reaction (AAR) in concretes presenting different strengths and incorporating a wide range of reactive aggregate types and natures, Cem. Concr. Res. 93 (2017) 17–31. https://doi.org/10.1016/j.cemconres.2016.12.001.

[8]   V.E. Saouma, Diagnosis & prognosis of AAR affected structures : state-of-the-art report of the RILEM Technical Committee 259-ISR, Springer, Cham, Switzerland, 2021.

[9]   M.A. Berube, B. Durand, D. Vézina, B. Fournier, Alkali-aggregate reactivity in Québec (Canada), Can. J. Civ. Eng. 27 (2000) 226–245. https://doi.org/10.1139/cjce-27-2-226.

[10]   R.N. Swamy, Assessment and Rehabilitation of AAR-affected Structures, Cem. Concr. Compos. 19 (1997) 427–440.

[11]   A. Shayan, A. Xu, F. Andrews-Phaedonos, Assessment of alkali-aggregate reaction in three bridges and its implications for maintenance of the structures, Road Transp. Res. 24 (2015) 54–66.

[12]   B.P. Gautam, D.K. Panesar, S.A. Sheikh, F.J. Vecchio, N. Orbovic, Alkali aggregate reaction in nuclear concrete structures: Part 2: Concrete materials aspects, 2015.

[13]   The Institution of Structural Engineers, Structural effects of alkali-silica reaction: Technical

guidance on the appraisal of existing structures, 1992.

[14]   G.E. Blight, M.G. Alexander, Alkali-aggregate reaction and structural damage to concrete : engineering assessment, repair and management, CRC Press/Balkema, Leiden, The Netherlands, 2011. https://doi.org/10.1201/b10773.

[15]   L.F.M. Sanchez, T. Drimalas, B. Fournier, D. Mitchell, J. Bastien, Comprehensive damage assessment in concrete affected by different internal swelling reaction (ISR) mechanisms, Cem. Concr. Res. 107 (2018) 284–303. https://doi.org/10.1016/j.cemconres.2018.02.017.

[16]   A. Zahedi, C. Trottier, L.F.M. Sanchez, M. Noël, Microscopic assessment of ASR-affected concrete under confinement conditions, Cem. Concr. Res. 145 (2021) 106456. https://doi.org/10.1016/j.cemconres.2021.106456.

[17]   L.F.M. Sanchez, B. Fournier, D. Mitchell, J. Bastien, Condition assessment of an ASR-affected overpass after nearly 50 years in service, Constr. Build. Mater. 236 (2020) 117554. https://doi.org/10.1016/j.conbuildmat.2019.117554.

[18]   M.D.A. Thomas, K.J. Folliard, B. Fournier, P. Rivard, T. Drimalas, S.I. Garber, Methods for Evaluating and Treating ASR-Affected Structures: Results of Field Application and Demonstration Projects. Volume II: Details of Field Applications and Analysis Final Report, Fed. Highw. Adm. FHWA-HIF-1 (2013) 338.

[19]   R. Rivard, Quantitative Petrographic Technique for Concrete Damage Due to ASR : Experimental and Application, (2000) 63–72.

[20]   M.D.A. Thomas, K.J. Folliard, B. Fournier, P. Rivard, T. Drimalas, Methods for Evaluating and Treating ASR-Affected Structures: Results of Field Application and Demonstration Projects—Volume II: Details of Field Applications and Analysis, Washington, DC, n.d.

[21]   P.C. Chang, A. Flatau, S.C. Liu, Review paper: health monitoring of civil engineering, Struct. Heal. Monit. 2 (2003) 257–267.

[22]   S. Kashif Ur Rehman, Z. Ibrahim, S.A. Memon, M. Jameel, Nondestructive test methods for concrete bridges: A review, Constr. Build. Mater. 107 (2016) 58–86. https://doi.org/10.1016/j.conbuildmat.2015.12.011.

[23]   B. Godart, P. Fasseu, M. Michel, Diagnosis and monitoring of concrete bridges damaged by AAR in Northern France, in: 9th Int. Conf. Alkali-Aggregate React. Concr. July 1992, London, Vol. 1, 1992.

[24]   N. Smaoui, B. Fournier, M.-A. Bérubé, B. Bissonnette, B. Durand, Evaluation of the expansion attained to date by concrete affected by alkali silica reaction. Part II: Application to nonreinforced concrete specimens exposed outside, Can. J. Civ. Eng. 31 (2004) 997–1011. https://doi.org/10.1139/l04-074.

[25]   M.A. Bérubé, N. Smaoui, B. Fournier, B. Bissonnette, B. Durand, Evaluation of the expansion attained to date by concrete affected by ASR - Part III: Application to existing

structures, Can. J. Civ. Eng. 32 (2005) 463–479.

[26] D.J. Deschenes, O. Bayrak, K. Folliard, ASR/DEF-damaged bent caps:shear tests and field implications, University of Texas, 2009.

[27] ASTM C856-20, Standard Practice for Petrographic Examination of Hardened Concrete, Annu. B. ASTM Stand. i (2004) 1–17. https://doi.org/10.1520/C0856.

[28] P.E. Grattan-Bellew, L.D. Mitchell, Quantitative petrographic analysis of concrete: the damage rating index (DRI) method, a review, (2006).

[29] V. Villeneuve, B. Fournier, J. Duchesne, Determination of the damage in concrete affected by ASR- the damage rating index (DRI), in: Proc. 14th Int. Conf. Alkali-Aggregate React. Concr., 2012.

[30] V. Villeneuve, Détermination de l'endommagement du béton par méthode pétrographique quantitative, M.Sc., Université Laval, 2011.

[31] L.F.M. Sanchez, B. Fournier, M. Jolin, M.A.B. Bedoya, J. Bastien, J. Duchesne, Use of Damage Rating Index to quantify alkali-silica reaction damage in concrete: Fine versus coarse aggregate, ACI Mater. J. 113 (2016) 395–407. https://doi.org/10.14359/51688983.

[32] L.F.M. Sanchez, B. Fournier, M. Jolin, J. Bastien, Evaluation of the Stiffness Damage Test (SDT) as a tool for assessing damage in concrete due to alkali-silica reaction (ASR): Input parameters and variability of the test responses, Constr. Build. Mater. 77 (2015) 20–32. https://doi.org/10.1016/j.conbuildmat.2014.11.071.

[33] L.F.M. Sanchez, B. Fournier, M. Jolin, J. Bastien, D. Mitchell, Practical use of the Stiffness Damage Test (SDT) for assessing damage in concrete infrastructure affected by alkali-silica reaction, Constr. Build. Mater. 125 (2016) 1178–1188. https://doi.org/10.1016/j.conbuildmat.2016.08.101.

[34] L.F.M. Sanchez, B. Fournier, M. Jolin, J. Bastien, Evaluation of the stiffness damage test (SDT) as a tool for assessing damage in concrete due to ASR: Test loading and output responses for concretes incorporating fine or coarse reactive aggregates, Cem. Concr. Res. 56 (2014) 213–229. https://doi.org/10.1016/j.cemconres.2013.11.003.

[35] J.B. Walsh, The effect of cracks on the uniaxial elastic compression of rocks, J. Geophys. Res. 70 (1965) 399–411.

[36] R.S. Crouch, J.G.M. Wood, Damage evolution in AAR affected concretes, Eng. Fract. Mech. 35 (1990) 211–218. https://doi.org/10.1016/0013-7944(90)90199-Q.

[37] T.M. Chrisp, P. Waldron, J.G.M. Wood, Development of a non-destructive test to quantify damage in deteriorated concrete, Mag. Concr. Res. 45 (1993) 247–256.

[38] T.M. Chrisp, J.G.M. Wood, P. Norris, Towards quantification of microstructural damage in AAR deteriorated concrete, Fract. Concr. Rock, Recent Dev. (1989).

[39] N. Smaoui, B. Bissonnette, M.A. Bérubé, B. Fournier, Stresses induced by alkali-silica reactivity in prototypes of reinforced concrete columns incorporating various types of reactive aggregates, Can. J. Civ. Eng. 34 (2007) 1554–1566. https://doi.org/10.1139/L07-063.

[40] Y. Zhu, A. Zahedi, L.F.M. Sanchez, B. Fournier, S. Beauchemin, Overall assessment of alkali-silica reaction affected recycled concrete aggregate mixtures derived from construction and demolition waste, Cem. Concr. Res. 142 (2021) 106350. https://doi.org/10.1016/j.cemconres.2020.106350.

[41] B. Fournier, M.. Berube, K.. Folliard, M. Thomas, Report on the Diagnosis, Prognosis, and Mitigation of Alkali-Silica Reaction (ASR) in Transportation Structures, n.d.

[42] J. Boye Bork, C. Preben, Petrographic examination of hardened concrete, Bull. Int. Assoc. Eng. Geol. 39 (1989) 99–103. https://doi.org/10.1007/BF02592541.

[43] L.F.M. Sanchez, B. Fournier, M. Jolin, J. Duchesne, Reliable quantification of AAR damage through assessment of the Damage Rating Index (DRI), Cem. Concr. Res. 67 (2015) 74–92. https://doi.org/10.1016/j.cemconres.2014.08.002.

[44] Standard counsil of Canada, CSA23.2-9C: Compressive strength of cylindrical concrete specimens, in: CSA A23.119/CSA A23.219 Natl. Stand. Canada, 2019: pp. 733–748.

[45] Standard counsil of Canada, CSA23.2-14C: Obtaining and testing drilled cores for compressive strength testing, in: CSA A23.119/CSA A23.219 Natl. Stand. Canada, 2019: pp. 774–778.

[46] J. Tanesi, R. Meininger, Freeze-thaw resistance of concrete with marginal air content, Transp. Res. Rec. (2007). https://doi.org/10.3141/2020-08.

[47] F. Gong, Y. Takahashi, I. Segawa, K. Maekawa, Mechanical properties of concrete with smeared cracking by alkali-silica reaction and freeze-thaw cycles, Cem. Concr. Compos. 111 (2020) 103623. https://doi.org/10.1016/j.cemconcomp.2020.103623.

[48] M.A. Bérubé, D. Chouinard, M. Pigeon, J. Frenette, M. Rivest, D. Vézina, Effectiveness of sealers in counteracting alkali-silica reaction in highway median barriers exposed to wetting and drying, freezing and thawing, and deicing salt, Can. J. Civ. Eng. 29 (2002) 329–337. https://doi.org/10.1139/l02-010.

[49] G. Giaccio, R. Zerbino, J.M. Ponce, O.R. Batic, Mechanical behavior of concretes damaged by alkali-silica reaction, Cem. Concr. Res. 38 (2008) 993–1004. https://doi.org/10.1016/j.cemconres.2008.02.009.

[50] A. Mohammadi, E. Ghiasvand, M. Nili, Relation between mechanical properties of concrete and alkali-silica reaction (ASR); a review, Constr. Build. Mater. 258 (2020) 119567. https://doi.org/10.1016/j.conbuildmat.2020.119567.

[51] H. Aryan, B. Gencturk, M. Hanifehzadeh, J. Wei, ASR Degradation and Expansion of Plain

and Reinforced Concrete, Struct. Congr. 2020 - Sel. Pap. from Struct. Congr. 2020. (2020) 303–315. https://doi.org/10.1061/9780784482896.029.

[52] T. Garcia, M.S. Mirza, Effect of reinforcement on aar expansion in concrete, Proc. Int. Conf. Appl. Codes, Des. Regul. (2005) 271–279.

[53] A. Allard, S. Bilodeau, F. Pissot, B. Fournier, J. Bastien, B. Bissonnette, Expansive behavior of thick concrete slabs affected by alkali-silica reaction (ASR), Constr. Build. Mater. 171 (2018) 421–436. https://doi.org/10.1016/j.conbuildmat.2018.03.159.

[54] S. Multon, F. Toutlemonde, Effect of applied stresses on alkali-silica reaction-induced expansions, Cem. Concr. Res. 36 (2006) 912–920. https://doi.org/10.1016/j.cemconres.2005.11.012.

[55] B.P. Gautam, D.K. Panesar, S.A. Sheikh, F.J. Vecchio, Multiaxial expansion-stress relationship for alkali silica reaction-affected concrete, ACI Mater. J. 114 (2017) 171–184. https://doi.org/10.14359/51689490.

[56] J. Liaudat, I. Carol, C.M. López, V.E. Saouma, ASR expansions in concrete under triaxial confinement, Cem. Concr. Compos. 86 (2018) 160–170. https://doi.org/10.1016/j.cemconcomp.2017.10.010.

[57] A.E.K. Jones, L.A. Clark, The practicalities and theory of using crack width summation to estimate ASR expansion, Proc. Inst. Civ. Eng. Struct. Build. 104 (1994) 183–192. https://doi.org/10.1680/istbu.1994.26327.

[58] N. Smaoui, M.A. Bérubé, B. Fournier, B. Bissonnette, B. Durand, Evaluation of the expansion attained to date by concrete affected by alkali-silica reaction. Part I: Experimental study, Can. J. Civ. Eng. (2004). https://doi.org/10.1139/L04-051.

[59] M.R. Sakr, M.T. Bassuoni, Performance of concrete under accelerated physical salt attack and carbonation, Cem. Concr. Res. 141 (2021) 106324. https://doi.org/10.1016/j.cemconres.2020.106324.

[60] M.T. Bassuoni, M.M. Rahman, Response of concrete to accelerated physical salt attack exposure, Cem. Concr. Res. 79 (2016) 395–408. https://doi.org/10.1016/j.cemconres.2015.02.006.

[61] M.R. Sakr, M.T. Bassuoni, R.D. Hooton, T. Drimalas, H. Haynes, K.J. Folliard, Physical salt attack on concrete: Mechanisms, influential factors, and protection, ACI Mater. J. 117 (2020) 253–268. https://doi.org/10.14359/51727015.

[62] H. Haynes, M.T. Bassuoni, Physical Salt Attack on Concrete, Concr. Int. 33 (2011) 38–42.

[63] T.C. Powers, Freezing effects in concrete, Durab. Concr. (1975). https://doi.org/10.14359/17603.

[64] A Working Hypothesis for Further Studies of Frost Resistance of Concrete, ACI J. Proc.

(1945). https://doi.org/10.14359/8684.

[65]   R.A. Deschenes, E.R. Giannini, T. Drimalas, B. Fournier, W.M. Hale, Mitigating alkali-silica reaction and freezing and thawing in concrete pavement by silane treatment, ACI Mater. J. 115 (2018) 685–694. https://doi.org/10.14359/51702345.

[66]   M.A. Bérubé, D. Chouinard, M. Pigeon, J. Frenette, L. Boisvert, M. Rivest, Effectiveness of sealers in counteracting alkali-silica reaction in plain and air-entrained laboratory concretes exposed to wetting and drying, freezing and thawing, and salt water, Can. J. Civ. Eng. 29 (2002) 289–300. https://doi.org/10.1139/l02-011.

[67]   R.A. Deschenes, E.R. Giannini, T. Drimalas, B. Fournier, W.M. Hale, Effects of moisture, temperature, and freezing and thawing on Alkali-Silica reaction, ACI Mater. J. 115 (2018) 575–584. https://doi.org/10.14359/5170219.

4.2   Introduction

Alkali-silica reaction (ASR) is one of the most harmful durability-related distress mechanisms in concrete. In the presence of moisture, uncrystallized/poorly crystallized mineral phases (i.e., silica) present in fine or coarse aggregates may react with the alkali hydroxides from the concrete pore solution to form an expansive secondary product, the so-called "ASR gel" [1]. This gel, swells upon water uptake from the surroundings, leading to induced expansion, cracking and reduction of mechanical properties of the affected material [1,2]. In concrete structures, cracking may extend to numerous locations and structural members (i.e., columns, beams, slabs, etc.), leading to a loss in serviceability (due to deformations) and structural capacity [3–5].

Decades of extensive research in the field of ASR have yielded a quite thorough understanding of the reaction mechanism, kinetics, mechanical implications, and preventive measures [1,6]. However, one of the most challenging tasks is the prediction of future expansion in affected structures (i.e., prognosis). Bérubé et al. [7] developed a procedure for the evaluation of 'potential residual expansion' through core-based evaluations in accelerated conditions (i.e., 38°C and 100% relative humidity – RH). Research conducted using this tool has shown quite promising results [8–11]. However, expansion obtained in laboratory conditions is prone to differences compared to expansion an affected structure due to vast distinctions between the two scenarios including exposure conditions (i.e., temperature and RH), confinement conditions (i.e., the presence of reinforcement or external restraints) and other factors that are challenging to replicate accurately in a laboratory environment [12,13]. Moreover, such tests are limited to accelerated conditions and thus do not provide an accurate timeline for the development of expansion in real structures.

Due to the complexity of predicting future ASR-induced expansion and deterioration in affected concrete, many researchers have resorted to micro and macro-scale computational models [12,14–21]. Such frameworks are often built upon proposed or modified existing semi-empirical, analytical, and or numerical expressions employed to predict the reaction kinetics in ASR-affected structures, typically based upon observations drawn from laboratory testing. One such model was the semi-empirical approach proposed by Larive (1998) [19] for the description of ASR kinetics and induced expansion in laboratory; yet, until recently it required calibration in order to give the semi-empirical expression a physical meaning [21]. Nguyen et al. [22] expanded on the work done by De Grazia et al. [21] by incorporating the effect of leaching and exposure conditions. However, this approach has never been validated using real structures due to unaddressed challenges in correlating laboratory and field expansion. Thus, the objective of this work is presenting further modifications to enable the validation and application of existing ASR expansion models to real structures.

4.3   Factors affecting the expansion of reinforced concrete in the field

Many researchers have shown that the mechanical performance of ASR-affected concrete is directly influenced by induced expansion. However, in such research, expansion is often regarded as a one-dimensional property, yet when considering structural members, anisotropy is crucial to accurately predict the mechanical performance of an affected member. In this work, 'anisotropy' refers to the asymmetrical expansion of ASR-affected concrete, resulting from non-intrinsic properties of affected concrete members (e.g., casting direction, member geometry, confinement conditions, etc.).

4.3.1   Geometry and casting direction

Researchers have shown that for a given ASR-affected concrete member, expansion is almost always greatest when measured in the direction perpendicular to the casting plane [23–28]. According to Smaoui et al. [23], this behaviour is dictated by 'flaky/elongated' aggregate particles which are distributed along the casting plane once the cast concrete settles, resulting in a greater area of reactive particles. Moreover, the resulting weak interfaces in the orthogonal plane generate less resistance to swelling pressures, and thus greater expansion in the vertical direction [23].

As part of the above study, Smoaui et al. [23] also calculated a coefficient of anisotropy for each of the specimens monitored which is calculated as the ratio between expansion perpendicular and parallel to the casting plane. According to the author's earlier argument on the effect of the casting plane, one may expect that for a "shorter" member, a greater number of weak interfaces could be present, resulting in higher expansion. Thus, by linking the coefficient of anisotropy with the relative geometry of the member (e.g., ratio of vertical length to longitudinal/transversal length), one could develop a relationship for the effect of geometry on anisotropic expansion. However, this requires further analysis of expansion data which has not been conducted by the authors.

4.3.2   Restraint conditions

Several research works have addressed the anisotropic effect of confinement on ASR expansive strains showing that confinement along a given direction reduces induced expansion along that same direction (i.e., confined expansion < free expansion) [25,26,28–33]. Over the years, researchers have examined the coefficient of anisotropy for confinement, defined as the ratio between confined and free expansion, to better understand this effect. Relationships between the coefficient of anisotropy and reinforcement ratio often portray this effect as a sudden drop followed by a steady plateau as shown in Figure 4.1A, regardless of the type or nature of reactive aggregate present [28,29,32]. According to the ISE [4], concrete expansion results in the development of local tensile stresses due to the naturally strong bond between hardened concrete and steel reinforcement. As a result, compressive stress develops in the concrete (Figure 4.1B), counteracting the induced expansion.

A) Expansion reduction due to reinforcement   B) Induced compressive stress due to reinforcement

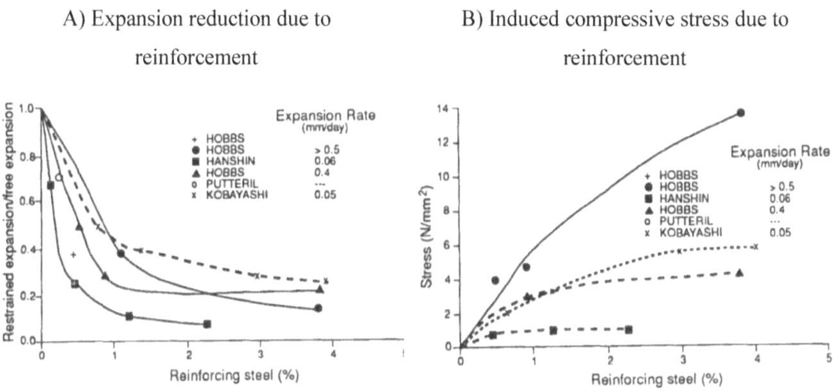

Figure 4.1: Effects of reinforcement on ASR free expansion and induced stress in concrete as per ISE (1992) [4].

According to the above explanation, it is plausible to state that mechanism by which confinement reduces expansion is quite similar for both reinforcement and compressive stresses, meaning that conclusions drawn from one mechanism could be applicable to both. However, contrary to the effect of reinforcement, which was often examined in singly reinforced members, the effect of multi-dimensional stress states on anisotropic ASR expansion has been studied by numerous researchers [25,30,31]. Interestingly, research conducted by Multon & Toutlemonde [25] indicated that, regardless of the stress state (i.e., radial, uniaxial, biaxial, or triaxial), volumetric expansion tends to remain constant. This is often referred to as 'expansion transfer' since confinement does not reduce volumetric free expansion, but rather reduces it along the most confined direction, while compensating for the loss in less confined directions. Expansion transfer is not limited to confining stresses as many studies monitoring the effect of transversal reinforcement (i.e., stirrups) often report higher values along that direction than free expansion [26,34]. However, Gautam et al. [31] and Liaudat et al. [30] have shown that this phenomenon is only applicable up to a certain volumetric stress (i.e., $\approx$ 2MPa), beyond which volumetric expansion is reduced as a function of the latter [30].

4.4 Forecasting unrestrained ASR expansion of field concrete

4.4.1 Semi-empirical model for predicting AAR expansion

Over the years, many semi-empirical models have been developed with the aim of predicting laboratory-induced ASR expansion, and its impact on the physical properties of affected concrete. One of the most widely used models, proposed by Larive, interprets ASR expansion as an exponential function of reaction kinetics and ultimate expansion (Figure 4.2). The model was developed according to a semi-empirical approach, based upon an extensive experimental campaign (i.e., over 600 specimens tested), to describe the induced expansive behaviour of unconfined concrete in the laboratory. Larive's equation (Eq. 4.1) is based on three parameters (i.e., characteristic time $-\tau_C$, latency time $-\tau_L$, and ultimate expansion $-\varepsilon^\infty$), which are only valid for a given set of exposure conditions (i.e., temperature and RH). More specifically, temperature may slow down or accelerate the expansion process, yet ultimate expansion is affected by the aggregate reactivity and primary factors (i.e., alkali content and moisture uptake).

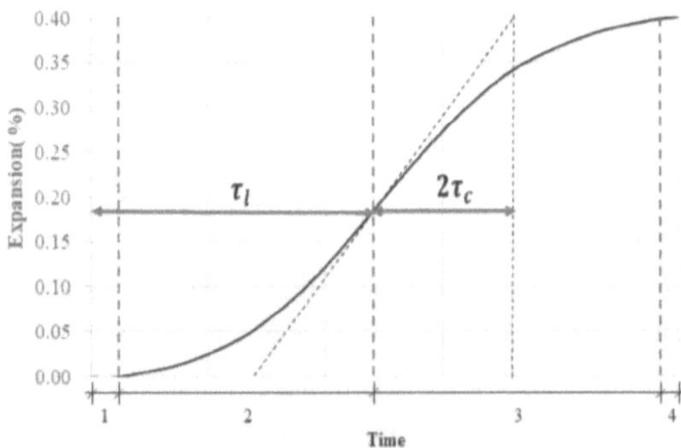

Figure 4.2: Interpretation of ASR expansion as per Larive's model [19,21].

$$\varepsilon(t,\theta) = \frac{1 - e^{-\frac{t}{\tau_C(\theta)}}}{1 + e^{-\frac{t-\tau_L(\theta)}{\tau_C(\theta)}}} \times \varepsilon^\infty \qquad (4.1)$$

Despite the ability of Larive's equation to demonstrate AAR induced expansion in affected laboratory specimens, the lack of physical meaning attributed to the equation's parameters (i.e., τ_C

& τ_L) implies that this approach is restricted in its ability to predict expansion as a function of material properties and/or exposure conditions. De Grazia et al. [21] addressed this gap through a rigorous experimental campaign incorporating different types and natures of reactive aggregates and mix designs so that Larive's parameters might be associated with the physical effects of AAR development for a wide range of conditions (i.e., alkali content, temperature, and RH). Consequently, a novel semi-empirical model (i.e., De Grazia's model – Eq. 4.2) was built, calibrated and validated to express AAR-induced expansion as a function of aggregate type and nature, alkali content, temperature and relative humidity.

$$\varepsilon(t, \theta) = \frac{1 - e^{-\frac{t}{\tau_C\, k_{C,T}\, k_{C,RH}\, k_{C,\%A}\, k_{C,E}}}}{1 + e^{-\frac{t - \tau_L k_{L,T}\, k_{L,RH}\, k_{L,\%A}\, k_{L,E}}{\tau_C\, k_{C,T}\, k_{C,RH}\, k_{C,\%A}\, k_{C,E}}}} \times (k_{Inf,T}\, k_{Inf,RH}\, k_{Inf,\%A}\, k_{Inf,E})\varepsilon^{\infty} \qquad (4.2)$$

One should note that despite the ability of De Grazia's model to predict ASR for a wide range of conditions, the proposed framework was developed using specimens in a laboratory environment (i.e., free expansion, constant temperature and RH). Thus, the direct use of this framework for the prediction of the ASR development of field structures/elements is premature as several conditions would be left unaccounted for (i.e., specimen geometry, orientation of casting direction, changes in exposure conditions, confinement effects).

4.4.2 Correlation between laboratory expansion and field expansion

Expanding on De Grazia's model, Nguyen et al. [22] developed a simplified framework to establish a correlation between laboratory observations and field expansion, while taking into consideration the influence of temperature, RH and alkali leaching from laboratory accelerated tests, which are critical in the development of ASR in structural elements. In the study, a two-phase procedure was developed to simulate the expansion curve for ASR-affected concrete in the field: 1) determining model parameters (i.e., ε^{∞}, τ_C and τ_L) while accounting for the effect of leaching; the resultant expansion curve will henceforth be referred to as the "ideal" expansion curve, and 2) the simulation of the ASR "free expansion" curve for the element under investigation through the consideration of exposure conditions (i.e., temperature and RH). Thus, Eq. 4.3, Eq. 4.4, and Eq. 4.5 describe the model parameters for expansion in field concrete, as proposed by Nguyen et al [22].

$$\varepsilon^{\infty} = \varepsilon^{\infty,0}(RA, T_0, RH_0, A_0) \cdot k_{\varepsilon,RH} \cdot k_{\varepsilon,A} \tag{4.3}$$

$$\tau_C = \tau_C^0(RA, T_0, RH_0, A_0) \cdot k_{C,T} \cdot \frac{1}{k_{C,LA}} \tag{4.4}$$

$$\tau_L = \tau_L^0(RA, T_0, RH_0, A_0) \cdot k_{L,T} \cdot \frac{1}{k_{L,LA}} \tag{4.5}$$

$\varepsilon^{\infty,0}$, τ_C^0 and τ_L^0 represent model parameters corresponding to a particular aggregate (RA) and cement alkali content (A_0) determined through laboratory accelerated testing conditions (i.e., temperature - $T_0 = 38°C$, relative humidity - $RH_0 = 100\%$). The effect of relative humidity on the ultimate expansion is accounted for through the coefficient $k_{\varepsilon,RH}$. Coefficients $k_{\varepsilon,A}$ and k_{ε,A_0} factor the effect of alkali leaching on expansion, whereas $k_{C,LA}$ and $k_{L,LA}$ describe the same effect on characteristic and latency times respectively. Finally, $k_{C,T}$, $k_{L,T}$ and $k_{\varepsilon,RH}$ are coefficients describing the effect of exposure conditions on ultimate ASR expansion and reaction kinetics. The subsequent sections will focus on introducing the model's considerations, defining the above coefficients, and outlining the two-phase calculation procedure mentioned above.

4.4.2.1 Determination of ideal expansion curve

Leaching takes place when dissolved alkali hydroxides are extracted from the concrete due a concentration gradient between internal and surface concrete resulting from very humid ambient conditions. As described by Kawabata [13], the above chemical process is quite prominent in accelerated testing conditions due to small sample sizes and high vapor pressure. On the other hand, structural elements exposed to unhindered exposure conditions do not experience any significant leaching due to conditions unlike those of laboratory specimens.

To establish a correlation between the alkali content of laboratory specimens and concrete in the field, it is important to lay out reasonable assumptions in order to help quantifying the above effect: 1) for a given concrete mix, the amount of leaching after one year (i.e., LA) in percentage (%) is assumed to be constant regardless of the initial alkali content (i.e., A₀) and, 2) the alkali content (i.e., A) required to compensate for leaching represents the alkali content for a scenario where no leaching occurs. The above assumptions are summarized in Eq. 4.6, where A and A₀ are expressed in kg/m³. Relationships in Eq. 4.7 – 4.9 represent the ideal expansion curve parameters.

$$A = \frac{A_0}{1 - LA} \tag{4.6}$$

$$\varepsilon^{\infty}_{no\ leaching} = \varepsilon^{\infty}_{leaching} \frac{k_{\varepsilon,A}}{k_{\varepsilon,A_0}} \tag{4.7}$$

$$\tau_{C,no\ leaching} = \tau_{C,leaching} \frac{1}{k_{C,LA}} \tag{4.8}$$

$$\tau_{L,no\ leaching} = \tau_{L,leaching} \frac{1}{k_{L,LA}} \tag{4.9}$$

<u>Alkali content dependency of ε^{∞}</u>

Experimental programs carried out by Fournier et al. [1], Shehata et al. [35], and Costa et al. [36], demonstrated significant variances in the ultimate expansion attained by specimens containing different alkali contents. To quantify the above effect, the coefficient of alkali content for ultimate expansion was defined as the ratio of ultimate expansion at a given alkali content to ultimate expansion for an alkali content of 5.25 kg/m³(i.e., $k_{\varepsilon,A}$). Based on the experimental data, a model (presented in Figure 4.3) was developed to describe this effect.

A)   B)

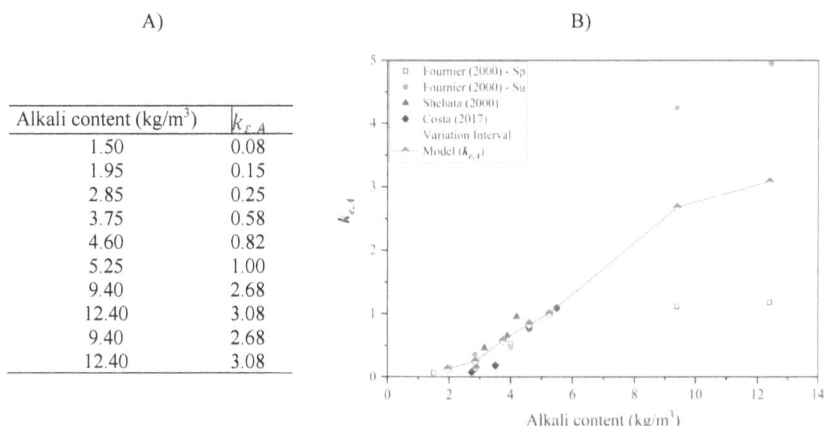

Alkali content (kg/m³)	$k_{\varepsilon,A}$
1.50	0.08
1.95	0.15
2.85	0.25
3.75	0.58
4.60	0.82
5.25	1.00
9.40	2.68
12.40	3.08
9.40	2.68
12.40	3.08

Figure 4.3: Proposed model for coefficient of alkali content adopted from [36], [35] and [1].

<u>Alkali leaching dependency of τ_C and τ_L</u>

Lindgård (2013) [37] and Shehata (2019) [27] monitored the accelerated expansion of concrete samples while recording leaching. Both authors reported different amounts of leaching based on varied laboratory induced conditions (i.e., using water or alkali solutions with different concentrations) and specimens of different sizes (70 x 70mm prisms and 100 x 100mm prisms from [37]). Combined with the expansion data of the aforementioned specimens, the effect of leaching on characteristic and latency times was determined (Figure 4.4). Based on the linear relationship obtained for both parameters, expressions in Eq. 4.10 and Eq. 4.11 were developed to describe each coefficient (i.e., $k_{C,LA}$ and $k_{L,LA}$) as a function of the amount of leaching (%) after one year.

A) Effect of alkali leaching on τ_C    B) Effect of alkali leaching on τ_L 

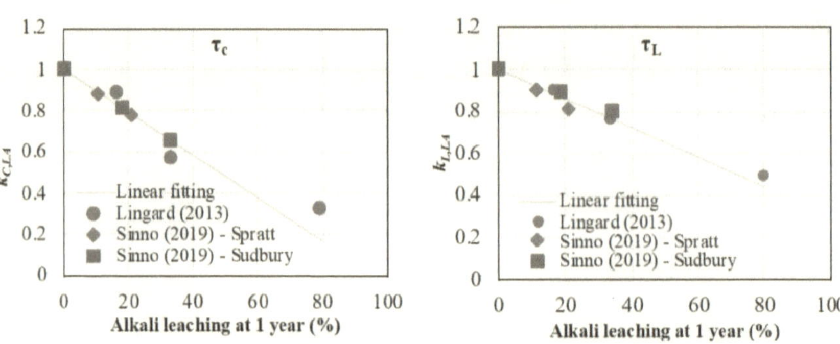

Figure 4.4: Deduced relations for alkali leaching coefficients for characteristic and latency times adopted from [37] and [38].

$$k_{C,LA} = 1 - 1.05LA \tag{4.10}$$

$$k_{C,LA} = 1 - 0.7LA \tag{4.11}$$

<u>Model validation</u>

The following validation was conducted using data from Sinno & Shehata (2019) [38]. The authors conducted ASR expansion testing on prismatic (70x70mm) and cylindrical (100x300mm) specimens with mixtures containing Spratt and Sudbury aggregates. The recorded leaching for the concrete made of Spratt aggregate after a year and a half was 25.8% for prisms and 13.5% for cylinders. Specimens prepared using the Sudbury aggregate achieved 41.3% and 22.7% of leaching for prisms and cylinders, respectively, after the same period of testing. It is important to mention that the leaching at one year was assumed to be 85% of the leaching determined after a year and a half (i.e., 22% for Spratt and 35% for Sudbury). The graphs in Figure 4.5 clearly indicate a greater gap between the ideal expansion curves and test data for the Sudbury aggregate, reflecting the higher leaching content. Moreover, the ideal expansion curves are closely matched for both types of specimens (i.e., prisms and cylinders) for each of the reactive aggregates, indicating that the specimen shape/geometry is not significantly influential on the simulated curve, as long as the leaching content is known.

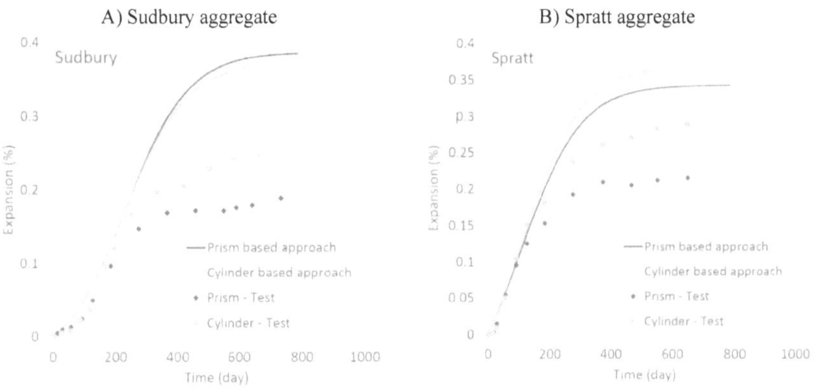

Figure 4.5: Experimental data and ideal expansion curves for specimens tested by [38].

4.4.2.2 Correlation with field expansion

Fluctuations in ambient temperature and RH could drastically influence expansion of concrete in the field (Figure 4.6). Thus, expanding on the work of Larive [19], Ulm et al. [20] evaluated the coupled phenomenon of ASR kinetics and heat diffusion on ASR expansion at the material and structural level. The authors observed a nearly perfect linear relationship between natural logarithms $ln(\tau_C)$ and $ln(\tau_L)$, separately, as a function of $1/T_i$ where T_i represents a given temperature. By reorganizing the proposed relationships, one could develop a relationship for normalized characteristic and latency times as a function of the laboratory and field temperatures (i.e., $T_0 = 38°C$ and T, respectively). Thus, Eq. 4.12 and Eq. 4.13 were proposed, where U_C and U_L represent the activation energy (i.e., threshold energy required to generate reaction) for characteristic and latency times, respectively. Moreover, Capra et al. [18] proposed a widely accepted model for normalized ultimate ASR expansion as a function of humidity (i.e., h), based on the principles of fracture mechanics (Eq. 4.14).

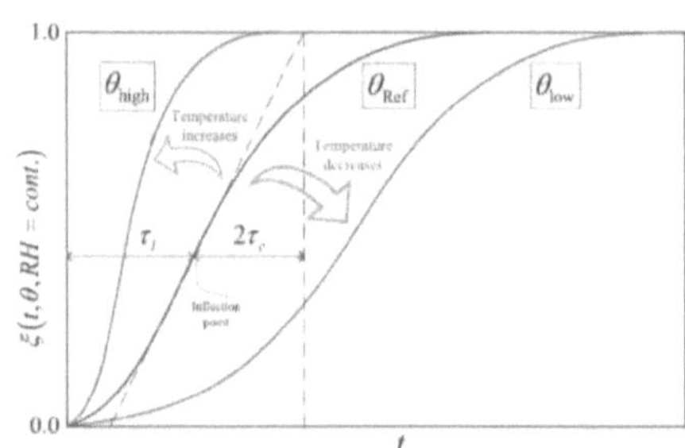

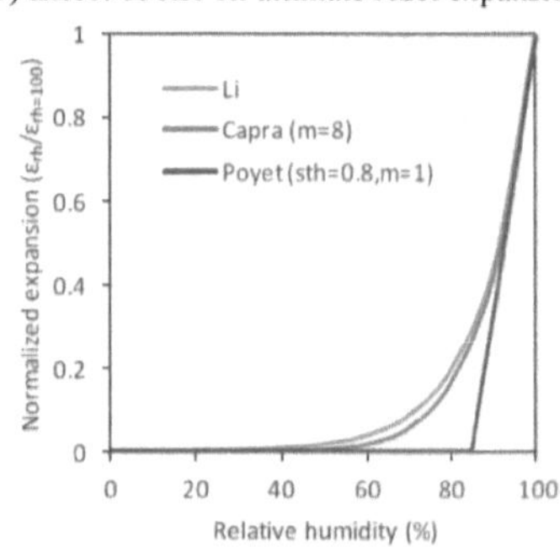

Figure 4.6: Influence of exposure conditions on ASR expansion [13,18,20].

$$k_{C,T} = \frac{\tau_C^T}{\tau_C^{T_0}} = \exp[U_C(\frac{1}{T} - \frac{1}{T_0})] \tag{4.12}$$

$$k_{L,T} = \frac{\tau_C^L}{\tau_C^{L_0}} = \exp[U_L(\frac{1}{T} - \frac{1}{T_0})] \tag{4.13}$$

$$k_{\varepsilon,RH} = \frac{\varepsilon^{RH=h}}{\varepsilon^{RH=100\%}} = h^8 \tag{4.14}$$

Finally, the timescale of periodic changes in exposure conditions must be considered. The following assumptions were made to simplify this process: 1) exposure conditions are assumed to fluctuate in monthly increments, and 2) exposure conditions are assumed to be the same on an annual basis. Nguyen et al. [22] adopted the model proposed by Kawabata et al. [13] for the calculation of changes in exposure conditions. Based on the flowchart displayed in Figure 4.7, every month is allocated a unique set of conditions (i.e., temperature and RH values), and thus a unique 'master curve'. Thus, changes in exposure conditions are reflected by 'moving' from the master curve of a given month to its counterpart for the following month as dictated by Figure 4.7.

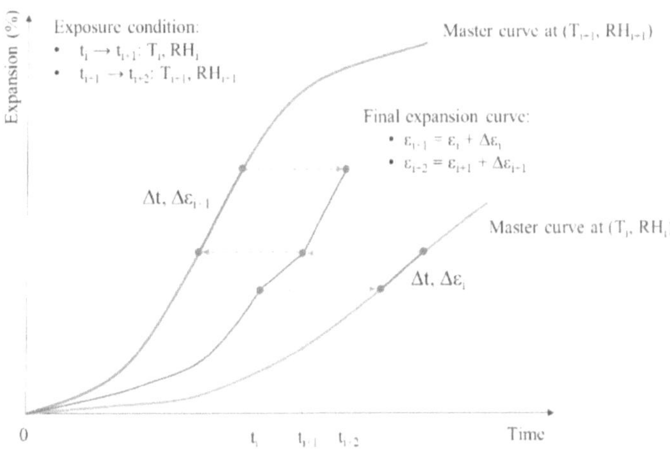

Figure 4.7: Schematic for calculation of temperature-dependent expansion as per [13] adopted from [22].

4.5 Semi-empirical approach for ASR expansion of field concrete under confinement

Factors affecting ASR-induced expansion (i.e., temperature, RH, alkali content) that were discussed have thus far been intrinsic properties of concrete, which are crucial in understanding the timescale and reaction kinetics of induced expansion on the material level. Anisotropy, on the other hand, is a non-intrinsic property of concrete, since it is not dependent on material properties, and thus only influences ultimate expansion.

4.5.1 Geometry and casting direction

Figure 4.8 represents labeled dimensions and expansions for rectangular and cylindrical concrete specimens to define relevant terminology to be used henceforth. As one may infer, $d_\perp$ and $\varepsilon_\perp$ refer to directions perpendicular to the casting plane whereas $d_\parallel$ and $\varepsilon_\parallel$ refer to the radial direction in cylindrical prisms. For rectangular prisms with unequal sides, $d_{\parallel,1}$, $\varepsilon_{\parallel,1}$, $d_{\parallel,2}$, and $\varepsilon_{\parallel,2}$ refer to directions parallel to the casting plane. Finally, Eq. 4.15 and Eq. 4.16 describes the average expansion for rectangular and cylindrical specimens, respectively.

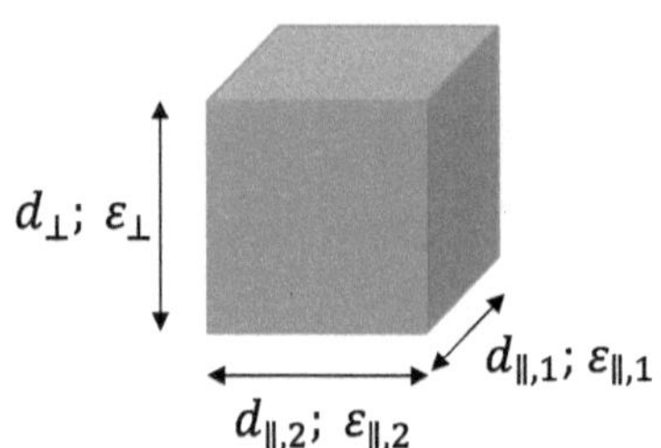 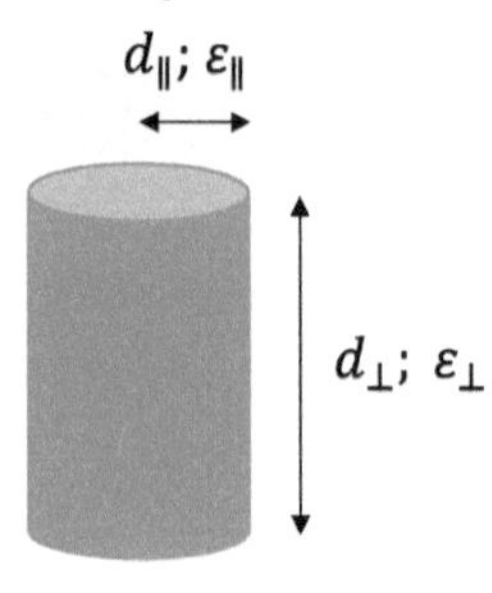

Figure 4.8: Labeled schematic of 3D concrete element.

$$\varepsilon_{avg} = \frac{\varepsilon_\perp + \varepsilon_{\parallel,1} + \varepsilon_{\parallel,2}}{3} \tag{4.15}$$

$$\varepsilon_{avg} = \frac{\varepsilon_\perp + 2\varepsilon_\parallel}{3} \tag{4.16}$$

To capture the effect of anisotropy in field expansion based on the model parameters of test specimens, it is essential to establish a link between one-dimensional expansion (measured in test specimens) and the three-dimensional expansion. Bearing this in mind, Eq. 4.17 – 4.19 describe anisotropic expansion in a given ASR-affected member, where ε_{avg}^∞ represents the average ultimate expansion determined in a test specimen. The above parameter can be determined using Eq. 4.15 and Figure 4.9A for concrete expansion prisms by assuming that $\varepsilon_{\parallel,1} = \varepsilon_{\parallel,2}$. For expansion cylinders, ε_{avg}^∞ can be determined using Eq. 4.16 and Figure 4.9B. Finally, α_1 and α_2 represent the geometry-dependent coefficients of anisotropy along each direction.

$$\varepsilon_\perp = \frac{3\varepsilon_{avg}^\infty}{\alpha_1 + \alpha_2 + 1} \tag{4.17}$$

$$\varepsilon_{\parallel,1} = \varepsilon_\perp / \alpha_1 \tag{4.18}$$

$$\varepsilon_{\parallel,2} = \varepsilon_\perp / \alpha_2 \tag{4.19}$$

The effect of the casting direction on expansion was investigated using expansion data collected by Smaoui et al. [23] and Giannini et al. [26]. Thus, Figure 4.9A and B represent the relation between expansion perpendicular and parallel to the casting plane for rectangular and cylindrical specimens, respectively, containing aggregates of different types (i.e., coarse and fine) and natures.

A) Rectangular specimens   B) Cylindrical specimens

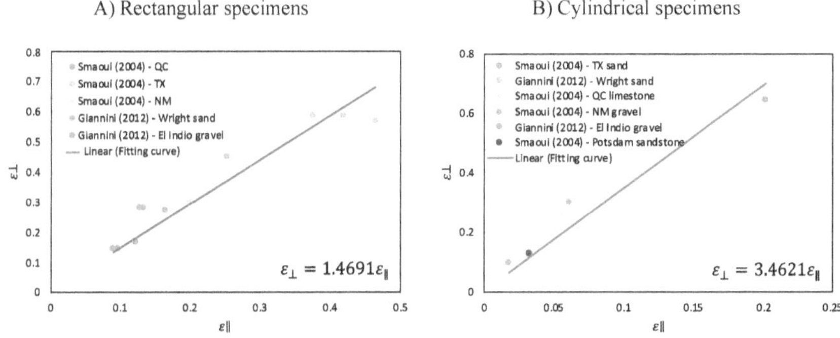

$$\varepsilon_\perp = 1.4691\varepsilon_\parallel \qquad \varepsilon_\perp = 3.4621\varepsilon_\parallel$$

Figure 4.9: Effect of casting direction [23,26].

According to Smaoui et al. [23], the relationships in Figure 4.9 are caused by weak interfaces in the vertical direction due to the heterogeneous distribution of aggregates. Thus, it seems logical to predict that this heterogeneity is somewhat governed by the ratio of $d_\perp/d_\parallel$. Indeed, the relationship between $\varepsilon_\perp/\varepsilon_\parallel$ and $d_\parallel/d_\perp$ (Figure 4.10) shows the existence of a linear trend. Thus, Eq. 4.20 describes the coefficient of anisotropy for member geometry.

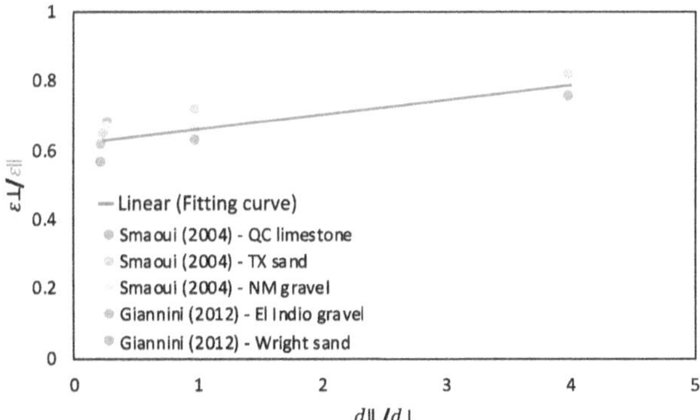

Figure 4.10: Effect of geometry and casting direction for rectangular specimens [23,26].

$$\alpha = \frac{\varepsilon_\perp}{\varepsilon_\parallel} = \frac{4}{97}\frac{d_\perp}{d_\parallel} + \frac{31}{50} \tag{4.20}$$

4.5.2 Confined anisotropic expansion

4.5.2.1 Reinforcement confinement

In the presence of reinforcement, induced expansion tends to be reduced in the more confined direction, whereas the difference is compensated by an increase in less confined directions. Generally, the more confined direction tends to possess longitudinal reinforcement (and thus greater steel reinforcement) whereas the less confined section contains stirrups. Thus, one may define the coefficients of anisotropy for more confined and less confined directions according to Eq. 4.21 and Eq. 4.22 / Eq. 4.23, respectively.

$$\delta_R = k_{\varepsilon,R} \tag{4.21}$$

$$\delta_{R_0,1} = k_{\varepsilon,R_0,1}\left(\frac{1 - k_{\varepsilon,R}}{2} * \frac{d_{R_0,2}}{average(d_{R_0,1}, \, d_{R_0,2})} + 1\right) \tag{4.22}$$

$$\delta_{R_0,2} = k_{\varepsilon,R_0,2}\left(\frac{1 - k_{\varepsilon,R}}{2} * \frac{d_{R_0,1}}{average(d_{R_0,1}, \, d_{R_0,2})} + 1\right) \tag{4.23}$$

The terms δ_R and δ_{R_0} represent the coefficients of anisotropy for reinforcement confinement for the more confined and less confined directions, respectively, where $k_{\varepsilon,R}$ and k_{ε,R_0} represent corresponding normalized confined expansion (i.e., confined expansion/free expansion) developed using existing expansion data (Figure 4.11) [26,28,29,39]. The dimension length terms (i.e., $d_{R_0,1}$ and $d_{R_0,2}$) are also considered to govern the extent of expansion transfer.

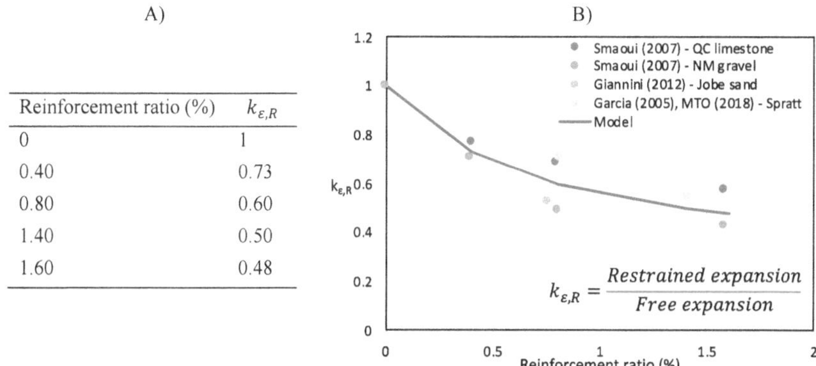

Figure 4.11: Effect of reinforcement on ASR expansion adopted from [28], [26], [29], and [39].

4.5.2.2 Stress confinement

The transfer effect due to stress confinement was developed according to the same concept as reinforcement. Thus, Eq. 4.24 and Eq. 4.25/ Eq. 4.26 represent the coefficients of anisotropy for more confined and less confined directions, respectively.

$$\delta_S = k_{\varepsilon,S} \tag{4.24}$$

$$\delta_{S_0,1} = \left(\frac{33.3 - pe_{un}^{fr}}{2} * \frac{d_{S_0,2}}{average(d_{S_0,1},\ d_{S_0,2})} + 100\right) \tag{4.25}$$

$$\delta_{S_0,2} = \left(\frac{33.3 - pe_{un}^{fr}}{2} * \frac{d_{S_0,1}}{average(d_{S_0,1},\ d_{S_0,2})} + 100\right) \tag{4.26}$$

The terms δ_S and δ_{S_0} indicate the expansion transfer coefficients of stress confinement for the more confined and less confined directions, respectively. Furthermore, $k_{\varepsilon,S}$ represents the coefficient for the effect of stress confinement corresponding to an induced expansion of pe_{un}^{fr}, which represents expansion along a confined direction as a percentage of volumetric free expansion (Eq. 4.27 and Eq. 4.28). It is important to note that the above equations are only applicable for uniaxial stress confinement, since accounting for expansion transfer under multiaxial stresses is quite complex and requires a more advanced mathematical approach.

$$pe_{un} = 11.4 + \frac{21.9}{1 + \left(\frac{f}{1.8}\right)^{6.5}} \; for \; f \geq 0 \tag{4.27}$$

$$k_{\varepsilon,S} = \frac{pe_{un}^{fr}}{pe_{un}^{f=0}} = \frac{pe_{un}^{fr}}{33.3} \tag{4.28}$$

4.5.3 Model formulation

Based on the proposed coefficients of anisotropy, one may determine the ultimate confined anisotropic expansion as per Eq. 4.29 – 4.31, followed by corresponding expansion curves based on Larive's equation. Thus, the modeling process for correlating laboratory and field expansion could be summarized as follows: 1) The ideal expansion curve is determined using the expansion data for the average expansion of relevant specimens in CPT conditions (i.e., 38°C and 100% RH), 2) A correlation is established between laboratory and field expansion by calibrating exposure conditions (i.e., temperature and RH) to determine the average field expansion, and 3) Coefficients of anisotropy for geometry and anisotropy are calculated to determine the confined expansion of an ASR-affected member.

$$\varepsilon_{\perp}^{\infty,C} = \varepsilon_{\perp} \times \delta_{R,\perp} \times \delta_{S,\perp} \tag{4.29}$$

$$\varepsilon_{\parallel,1}^{\infty,C} = \varepsilon_{\parallel,1} \times \delta_{R,\parallel,1} \times \delta_{S,\parallel,1} \tag{4.30}$$

$$\varepsilon_{\parallel,2}^{\infty,C} = \varepsilon_{\parallel,2} \times \delta_{R,\parallel,2} \times \delta_{S,\parallel,2} \tag{4.31}$$

Table 4.1 represents the input parameters considered for the laboratory specimens and the affected field member. The calculation procedure for the model could be broken down as follows:

1. Select accelerated expansion test data (i.e., CPT at 38°C and 100% RH) for concrete with similar strength and reactive aggregate as the affected member. Specify A_0 and LA for laboratory specimens.

2. Based on laboratory specimens' shape (i.e., cylinders or prisms), convert directional expansion data to average expansion data ε_{avg}^0 using models for the effect of geometry (i.e., Figure 4.9 B for cylinders or Figure 4.10 for prisms). Through trial and error, determine the best fitting curve parameters (i.e., $\varepsilon_{avg}^{\infty,0}$; τ_C^0; τ_L^0) for Larive's equation based on the average expansion data.

3. Specify the alkali content A, 12-month average temperature data T, and 12-month average relative humidity data RH for field member. Convert laboratory average expansion parameters to field average expansion parameters based on Eq. 4.12, 4.13, 4.14, and Figure 4.7.

4. Specify geometrical proportions for the field member $d_\perp$; $d_{\parallel,1}$; $d_{\parallel,2}$. Based on the model for the effect of geometry (i.e., Figure 4.10), determine the anisotropic ultimate expansion $\varepsilon_\perp^\infty$; $\varepsilon_{\parallel,1}^\infty$; $\varepsilon_{\parallel,2}^\infty$.

5. Specify reinforcement ratios $\rho_\perp$; $\rho_{\parallel,1}$; $\rho_{\parallel,2}$ and compressive axial stress f_r for field member. Determine confined anisotropic ultimate expansion $\varepsilon_\perp^{\infty,C}$; $\varepsilon_{\parallel,1}^{\infty,C}$; $\varepsilon_{\parallel,2}^{\infty,C}$ based on Eq. 4.29, 4.30, and 4.31.

6. Determine expansion curves for each direction $\varepsilon_\perp^C$; $\varepsilon_{\parallel,1}^C$; $\varepsilon_{\parallel,2}^C$ based on Larive's equation (i.e., Eq. 4.1).

Table 4.1: Input parameters for proposed ASR expansion model.

Element	Parameter	Notation
Laboratory specimens	Average ultimate expansion	$\varepsilon_{avg}^{\infty,0}$
	Characteristic time	τ_C^0
	Latency time	τ_L^0
	Alkali leaching	LA
	Initial alkali content	A_0
Field member	Geometric proportions	$d_\perp$; $d_{\parallel,1}$; $d_{\parallel,2}$
	Alkali content	A
	12-month average temperature	T
	12-month average relative humidity	RH
	Reinforcement ratios	$\rho_\perp$; $\rho_{\parallel,1}$; $\rho_{\parallel,2}$
	Compressive axial stress	f_r

It is important to emphasize that the main objective of the proposed model is predicting ASR expansion with the aim of selecting the most effective option for rehabilitation works, and not performing detailed structural modeling for the affected structure. Consequently, one could argue that this is a simplified model and thus appropriate assumptions must be considered. The above could be stated as follows:

- The damaged structure is only affected by ASR (i.e., effect of shrinkage, freezing and thawing, sulphate attack, etc. are not considered).

- No alkali leaching is present in the affected structure, since structural members are often too large for leaching to considerably affect expansion. Moreover, exposure conditions are not as extreme as accelerated conditions in the laboratory.

- Deformation due to creep is assumed to be zero, since no modeling considerations were incorporated for its calculation.

- Reinforcement and stress confinement are assumed to act independently of one another. Little to no research has been conducted on the influence of both types of confinement on each other, and thus, the relationship between the two is not clear.

- Expansion transfer is influenced by the level of confinement and geometric proportions of the affected member (i.e., greater expansion transfer along shorter direction). The latter was observed by Smaoui [23] and Giannini [26] and is considered in the expansion transfer equations for reinforcement and stress confinement.

- Simplified loading conditions are assumed for affected members (i.e., uniaxial loading only, no bending moment) since a more complex mathematical approach is required to account for multiaxial stresses as a function of the percentage of volumetric expansion.

- Uniaxial load is considered to be uniformly distributed across loading plane since finite element modeling is required for more detailed structural analysis.

Due to the simplified nature of the proposed model, some limitations may be expected. Ideally, the simulated expansion curves should represent the "worst-case" scenario; firstly, to confirm/eliminate the possibility of immediate structural concerns and secondly, to give and idea about required mitigation works. However, in the case of combined damage mechanisms (i.e., corrosion, FT, sulphate attack, etc.) expansion could be underestimated since the current model is unable to account for such scenarios. The same issue could occur when modeling affected members subjected to complex structural loadings (i.e., multi-directional loading, bending moments, flexure, etc.). Nevertheless, in such cases, modeled expansion curves could be potentially be used to perform more advanced structural modeling through finite element analysis.

4.6   Results and Discussion

4.6.1   Model validation

4.6.1.1 First validation: Kingston exposure site

Two distinct concrete mixtures containing Spratt reactive aggregate and prepared using low-alkali Portland cement (i.e., LAPC - 1.92kg/m3) and high-alkali Portland cement (i.e., HAPC - 3.33kg/m3) were used to cast reinforced (1.41%) and plain concrete beams with dimensions of 0.6 x 0.6 x 2m. The beams were left to expand in an ASR exposure site (Figure 4.12) in Kingston, Ontario (Canada) and were meticulously monitored over time (27 years). The average monthly temperature records of the city of Ottawa were used for this validation and were extracted from [40]. De Grazia's model was used to determine the best fitting curve for the average expansion of laboratory specimens casted using HAPC (Table 4.1 and Figure 4.13).

Figure 4.12: Beams from Kingston exposure site for ASR [41].

Table 4.2: Model parameters for HAPC mix.

Mixture	Initial alkali content (kg/m3)	τ_C	τ_L	ε^{∞}	1-year leaching (%)
HAPC	3.33	120	120	0.165	22

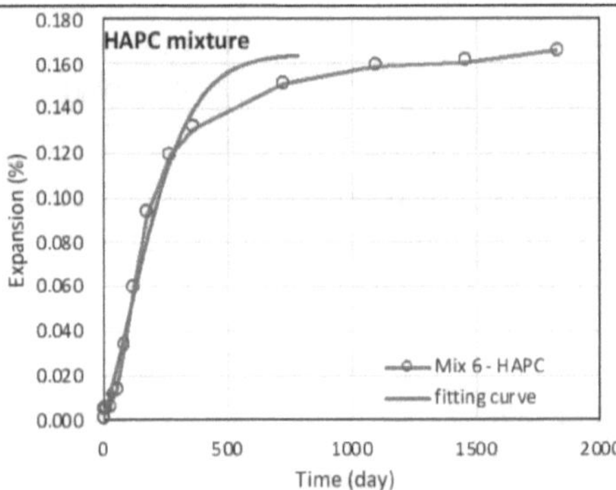

Figure 4.13: Fitting curve for expansion test results from HAPC mixture containing Spratt reactive aggregates [41].

Alkali-leaching was assumed to be 22% based on the works of Sinno & Shehata [38] using Spratt. Once the coefficients of anisotropy were calculated the longitudinal expansion of the beams was simulated. One may infer from Figure 4.14 that the simulation matches the readings from the field quite well for both mixtures, and for both free and confined scenarios. It is also worth noting that using model parameters from the HAPC mixture for LAPC mixtures worked quite well. The above indicates that the model is quite accurate and accommodating in terms of the fitting parameters, so long as the alkali content of the simulated member is known.

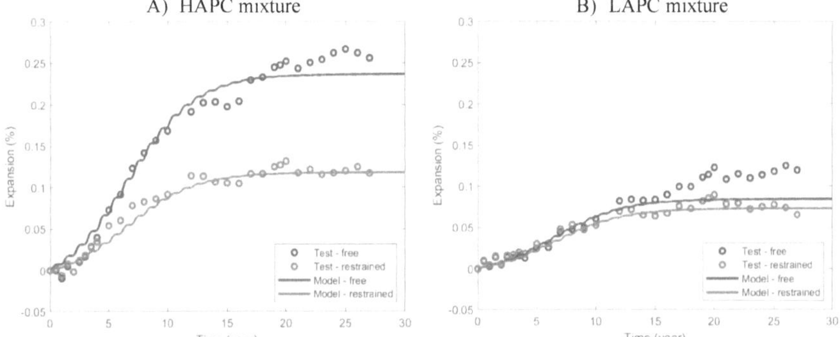

Figure 4.14: Field expansion from [41] and model simulation of 'free' expansion of Kingston site exposure beams.

4.6.1.2 Second validation: Texas exposure site

A concrete mixture containing Texas (TX) sand was prepared to cast plain and reinforced (i.e., 0.78% longitudinal and 0.21% stirrups) concrete blocks with dimensions of 380 x 380 x 810 mm [26]. Once again, the blocks were placed in an exposure site in Austin, Texas (USA) and left to expand. Frequent expansion readings were collected over a period of around 2 years. Contrary to the Kingston site beams, expansion was measured along all directions for members examined in this study (Figure 4.15). Thus, this presents an opportunity to further verify the coefficients of anisotropy developed for this novel model, yet the author reported observations in the form of average expansion of all three directions. The ratio of transversal expansion to longitudinal expansion was also recorded. Once again, average monthly temperature records for Texas was used from [40].

Figure 4.15: Locations of expansion measurements in exposure site blocks [26].

86

Based on the same procedure followed for the Kingston site beams, key information including model parameters (obtained from expansion testing), alkali content, alkali leaching, specimen geometry and reinforcement ratios were processed as model inputs. Expansion testing was conducted in laboratory conditions using cylinders [26]. Thus, this simulation also presents an opportunity to validate the deduced effect of casting direction for experimental cylindrical specimens (100mm x 200mm). Figure 4.16 and Table 4.2 demonstrate fitting curve parameters obtained for this mixture.

Table 4.3: Model parameters for TX sand.

Mixture	Initial alkali content (kg/m3)	τ_C	τ_L	ε^∞	1-year leaching (%)
Texas sand	5.25	50	40	0.400	35

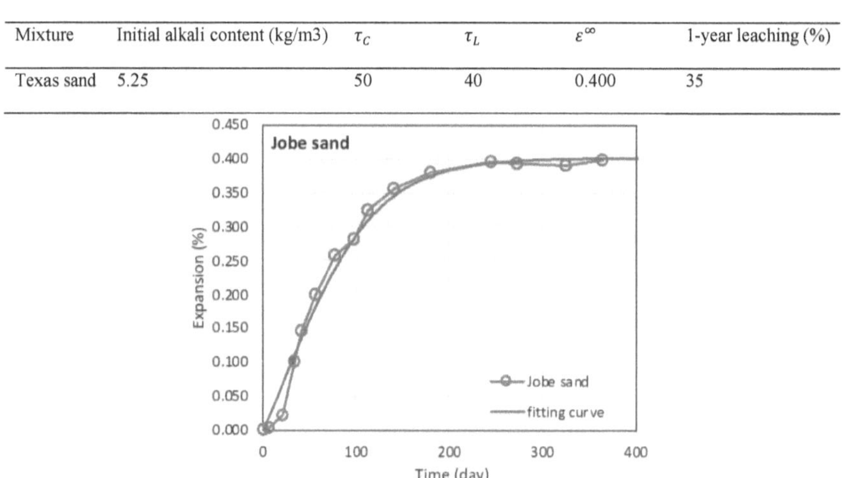

Figure 4.16: Fitting curve for TX sand [26].

Figure 4.17A (left and right) represents free and restrained expansion, respectively, of exposure site blocks cast using TX sand, a highly reactive fine aggregate. It is quite clear once again that the models for average expansion is in agreement with readings collected from the specimens for free and confined expansion. Furthermore, it is worth noting that the average expansion in both blocks remains virtually the same, yet expansion transfer from the restrained to less restrained directions may be observed. More specifically, longitudinal reinforcement appears to mitigate expansive strains along the longitudinal direction, whereas the presence of stirrups does not appear to be effective in mitigating transversal and vertical expansions. Finally, the ratio of transversal to longitudinal expansion was extracted from the model and compared with field readings in Figure

4.17B. It is important to note that the author defined transversal expansion as the average of the shorter lateral dimension and the vertical dimension. Results show that the coefficients of anisotropy for reinforcement are quite promising and are able to accurately predict anisotropy due to confinement.

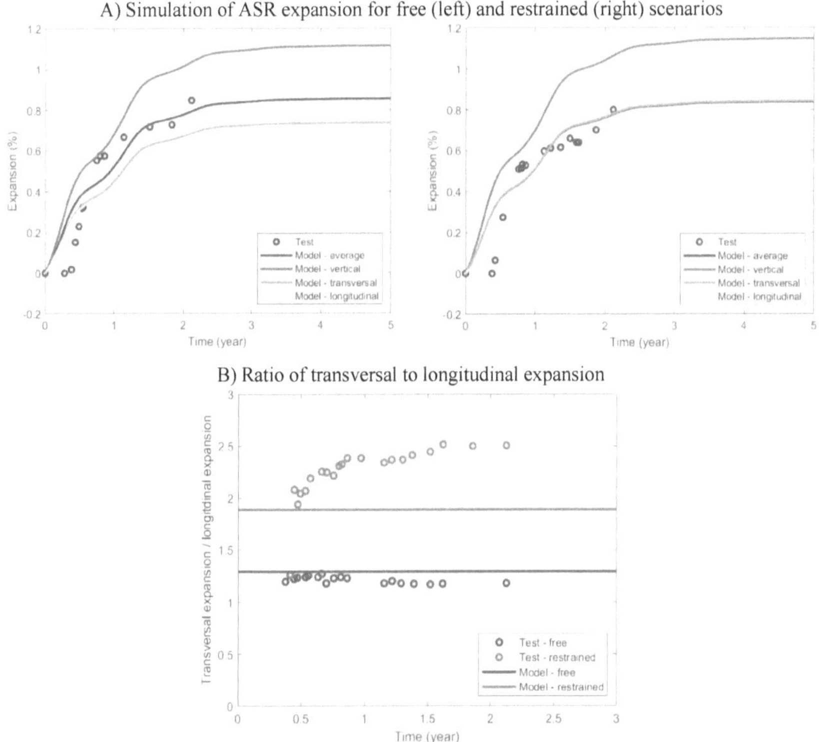

Figure 4.17: Model simulation of ASR expansion for exposure blocks cast with TX sand [26].

4.6.2   Case Study

4.6.2.1 S.I.T.E. building and its condition assessment

The School of Information Technology and Engineering (S.I.T.E.) building on the University of Ottawa campus (constructed in 2001) was inspected and found to portray surface signs indicating possible ASR damage. Based on the damage classification from visual inspection, four columns (i.e., two very-highly damaged and two highly damaged) were selected for coring. Laboratory investigations (microscopic and mechanical) conducted on core specimens confirmed the presence of ASR. Figure 4.18 displays photographs showing the damaged columns.

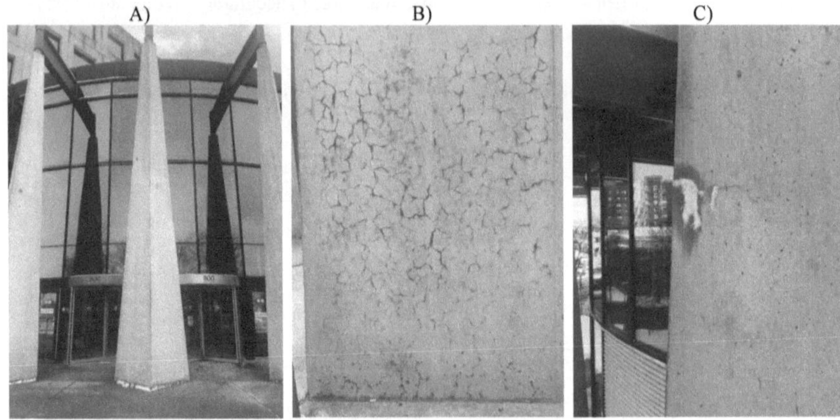

Figure 4.18: Photographs of columns affected by ASR.

Sanchez et al. [42,43] developed a diagnostic tool (i.e., multi-level assessment tool) for the appraisal of AAR-induced damage in affected concrete. The multi-level assessment tool is based upon a combination of microscopic and mechanical analyses (through the use of DRI and Stiffness Damage Test – SDT, respectively) and has been established through the appraisal of numerous concrete mixtures with various strengths (i.e., 25MPa, 35MPa, and 45MPa) incorporating 13 reactive aggregates with different lithotypes [42]. A data envelope was created after all the above analyses describing AAR-induced damage (i.e., expansion) as a function of microscopic and mechanical damage (Table 4.3). It is worth mentioning that the data was developed through an experimental program where test specimens were subject to 'free-expansion' conditions, whereas affected elements are usually subject to several restraints (i.e., reinforcing steel and/or service loads). Nevertheless, recent studies [34] revealed that the "overall" degree of ASR-induced damage may be appraised using the multi-level assessment tool, regardless of the confinement conditions.

Table 4.4: Proposed damage degrees based upon multi-level assessment.

Classification of ASR damage degree (%)	Reference expansion level (%)	Assessment of ASR				
		Stiffness loss (%)	Compressive strength loss (%)	Tensile strength loss (%)	SDI	DRI
Negligible	0.00 – 0.03	-	-	-	0.06 – 0.16	100 – 155
Marginal	0.04 ± 0.01	5 – 37	(-) 10 – 15	15 – 60	0.11 – 0.25	210 – 400
Moderate	0.11 ± 0.01	20 – 50	0 – 20	40 – 65	0.15 – 0.31	330 – 500
High	0.20 ± 0.01	35 – 60	13 – 25	45 – 80	0.19 – 0.32	500 – 765
Very high	0.30 ± 0.01	40 – 67	20 – 25		0.22 – 0.36	600 – 935

Figure 4.19 illustrates the potential expansion of the assessed columns based upon the multi-level assessment. Petrographic examination revealed that the reactive aggregate was a dolomitic limestone containing microcrystalline silica, with similar lithotype to Québec City limestone (QC). Thus, ASR damage was appraised according to the damage envelope of 45 MPa concrete incorporating QC. Column 4 was discovered to possess the greatest degree of damage (i.e., 0.08%), followed by column 3 (i.e., 0.05%). Finally, columns 1 and 7 were found to be the least damaged among all four columns (i.e., 0.03% for both). The importance of these results is such that they provide a reference as to whether expansion modelling results are reliable in their prediction of ASR development. In other words, simulation results should display an expansion between the range of 0.03% and 0.08% after almost 20 years.

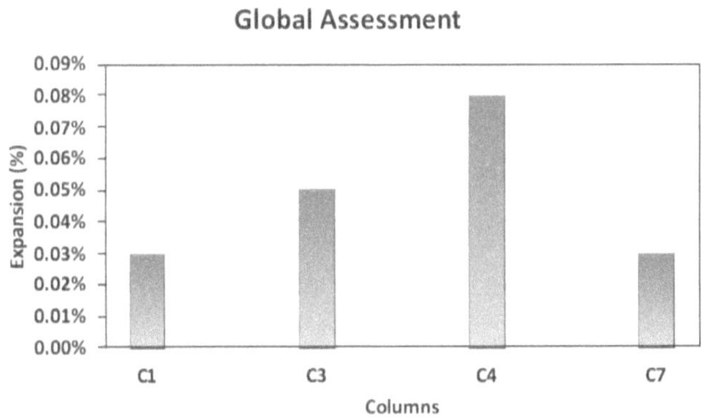

Figure 4.19: Potential expansion in assessed columns based upon multi-level assessment.

4.6.2.2 Simulation of free and confined anisotropic expansion

Before simulating the expansion of S.I.T.E. columns, it is quite important to address the lack of information due to inexistant construction records regarding the concrete mix used in the columns. Thus, the following assumptions were made for the reactive aggregate, cement alkali content, and leaching, respectively:

- To maintain consistency with the multi-level assessment of the columns, expansion data gathered from 25 MPa, 35 MPa, and 45 MPa QC mixes by [44] was used. The corresponding fitting curve parameters (Figure 4.20) were determined using De Grazia's model in CPT conditions (i.e., 38°C and 100% RH). Moreover, as per De Grazia et al., the alkali content was assumed to be 4.60 kg/m3.

- According to a survey on cements used in eastern Canada in the 1990s (around the time the S.I.T.E. building was constructed), an average alkali content of 0.87% was recorded [45]. Moreover, based on available literature, cement content used in Canadian construction in the year 2000 ranged between 325 kg/m3 and 450 kg/m3 to ensure that durability requirements are met [46]. Thus, the average value of 387.5 kg/m3 was assumed as the cement content used in the columns (i.e., alkali content = 0.87% x 387.5 = 3.37 kg/m3).

- Due to the lack of data provided by authors on leached alkalis, an assumption was made for leaching after one year. Sinno & Shehata [38] had previously observed leaching of 22% and 35% respectively after one year for Spratt and Sudbury aggregates respectively. Since Spratt and QC are both limestone-type aggregates, one-year leaching was presumed to be 22%.

- Internal temperature and RH were presumed to be the same as ambient conditions. Temperature was adapted from [40] based on average monthly records in Ottawa. Internal RH in the columns was assumed to be 100% in order to simulate the "worst-case scenario".

Table 4.5: Model parameters for S.I.T.E. columns.

Reactive aggregate	Initial alkali content (kg/m3)	τ_C	τ_L	ε^∞	1-year leaching (%)
QC	4.60	0	180	0.12	22

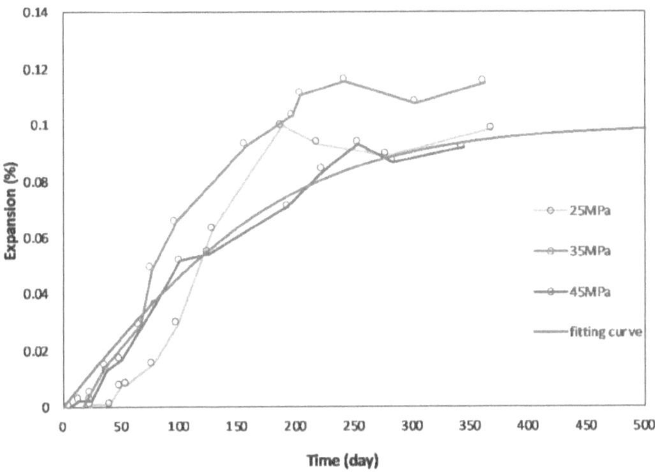

Figure 4.20: Fitting curve of expansion test results for QC limestone adopted from [44].

To compute the coefficients of anisotropy, it is important to consider the column's geometry and confinement conditions (i.e., reinforcement and stress). The columns are essentially pyramid-shaped with a height of 9.32 m, a minimum side length of 0.30 m (at the top), and a maximum side length of 1.25 m (at the bottom). The reinforcement content in the longitudinal direction is 0.74% and 0.20% for stirrups. Structural analysis software (i.e., SAP2000) was used to perform a simplified structural modeling based on loading conditions (i.e., dead and live loads) provided in engineering drawings; snow and wind loads were not considered in the analysis. Considering rigid restraints for all columns, the axial load (for a base cross section of 1.25 x 1.25m) and bending moment acting on each column were found to be 1.53MPa and zero, respectively. The latter indicates that the simplified structural approach adopted by the ASR model is quite suitable for this scenario since the column is only subjected to uniaxial compressive stress. A breakdown of the loads and modeling results are provided in Appendix A and Appendix B. The above information is summarized in Table 4.6.

Table 4.6: Column geometry and reinforcement.

Dimension	Notation	Length (mm)	Reinforcement ratio (%)	Compressive load (MPa)
Vertical	$d_\perp$	9320	0.78	1.53
Transversal - short	$d_{\parallel,S}$	300	0.21	0
Transversal - long	$d_{\parallel,L}$	1250	0.21	0

Figure 4.21 shows the simulated expansion of a single column in free expansion and confined conditions. The solid green and red curves represent the average transversal (i.e., average of $\varepsilon_{\parallel,S}$ and $\varepsilon_{\parallel,L}$) and vertical (i.e., $\varepsilon_\perp$) expansions, respectively, as labeled in Figure 4.21A and B and shown in Figure 4.21C, whereas the solid blue curve represents the average expansion of all three. Moreover, the dashed green curves represent the upper and lower bounds of transversal expansions labeled as $\varepsilon_{\parallel,S}$ and $\varepsilon_{\parallel,L}$, respectively. Despite changes in transversal and vertical expansions, the average expansion remains unchanged, indicating that confinement has no effect on volumetric expansion. On the other hand, confinement appears to mitigate expansion in the vertical direction, while slightly increasing expansion in the transversal direction. However, the resulting significant difference between transversal and vertical expansions could have important negative implications on the structural performance of the columns as ASR develops further in the column.

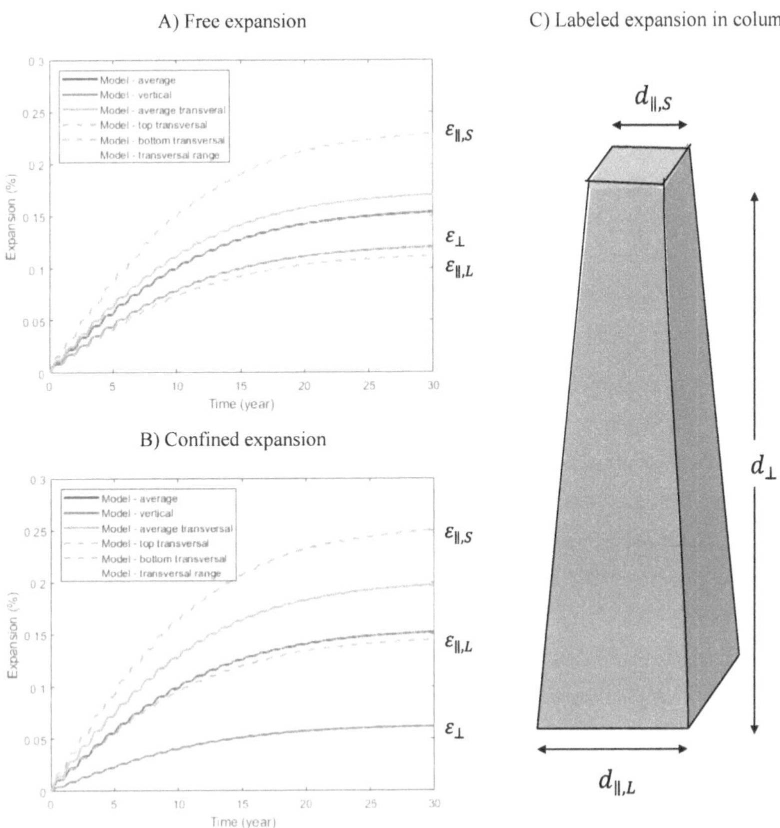

Figure 4.21: Simulation of ASR expansion in columns.

4.6.2.3 Prognosis of ASR-affected columns

To demonstrate and improve the prognostic potential of the developed model, the data analysis could be conducted to 1) use the model to predict future damage (i.e., expansion) based on condition assessment (i.e., diagnosis), and 2) combine the model with multi-level assessment to predict future reduction in mechanical properties. The latter could prove quite useful in cases where ASR damage may become a matter of concern with regards to structural performance (i.e., capacity to withstand service loads), since it has the potential to predict anisotropic mechanical properties of damaged concrete [5,47].

The thermodynamic coupling of temperature and RH for the modeling of expansion has been shown to produce quite accurate predictions based on validations conducted in this study. It is important to recall that this is achievable through the assumption that exposure conditions are constant throughout the simulated member. However, this is not the case in massive concrete members since the distribution of temperature and water is governed by diffusion mechanics in those scenarios, and thus detailed modeling of those members requires advanced approaches (i.e., finite element modeling software). Nevertheless, a simplified approach could be adopted to facilitate the prognosis of affected members by considering a small range of exposure conditions based on documented members in service with similar properties.

Thermal gradients in mass concrete members are described by [48] according to member depth (Figure 4.22). It is important to mention that the defined differentials are applicable for concrete systems with concrete decks according to CSA standards, yet the same data was successfully used in a similar fashion to model expansion in bridge piers [51]. Since the differentials in winter climates are influenced by the presence or absence of factors such as exposure to sun, wind, or precipitation, positive (+6°C) and negative (-6°C) differentials were both considered for winter conditions as per [48]. The average depth of S.I.T.E. columns was used to determine the differentials, which were then used to alter the monthly temperature records as per Table 4.6. It is important to indicate that based on the original monthly temperatures, the months of June, July, August, and September were considered as summer conditions, whereas the remaining months were considered to be winter conditions.

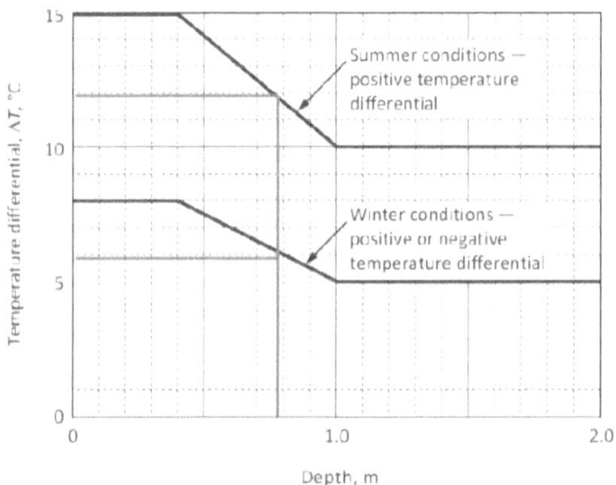

Figure 4.22: Temperature differentials as per [48].

Table 4.7: Upper and lower bounds for internal concrete temperature.

Month	Original average monthly temperatures (°C)	Lower bound (°C)	Upper bound (°C)
January	-10.8	-16.8	-4.8
February	-8.7	-14.7	-2.7
March	-2.5	-8.5	3.5
April	5.7	-0.3	11.7
May	13.4	7.4	19.4
June	18.3	30.3	30.3
July	20.9	32.9	32.9
August	19.5	31.5	31.5
September	14.3	26.3	26.3
October	7.8	1.8	13.8
November	1.0	-5.0	7.0
December	-7.1	-13.1	-1.1

According to some researchers, significant ASR expansion only occurs at an internal RH of 90% or greater, whereas others have reported that internal RH is often within the range of 90% to 95% in concrete structures [49,50]. However, based on the principles of heat diffusion, Gorga et al. [51] has calculated differentials in RH as a function of concrete depth, showing that the average RH in concrete columns is limited to 92% for climates in eastern Canada [19,20,51]. Thus, a range of 90 to 92% of RH was assumed for the modeling of S.I.T.E. columns. Based on the above discussions, upper (i.e., high temperature, high RH) and lower bounds (i.e., low temperature, low RH) were modeled for the columns in order to demonstrate long-term (i.e., over 50-year service life) and short-term (i.e., over the next 10 years) development of ASR damage.

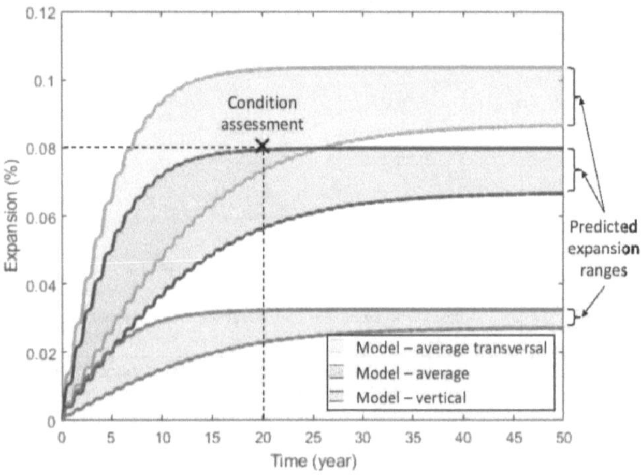

Figure 4.23: Predicted expansion of S.I.T.E. column.

Figure 4.23 demonstrates the prognosis of average expansion for S.I.T.E. columns over a traditional service life of 50 years, predicting a maximum average (i.e., blue band) expansion of ≈ 0.08% (i.e., moderate damage level based on multi-level assessment) in the short term (i.e., 10 years). Moreover, the maximum average transversal (i.e., green band) and vertical (i.e., red band) expansions of ≈ 0.10% and ≈ 0.03% are predicted in the same period of time. However, contrary to the above, the average expansion is a non-measurable quantity and thus is not practical in terms of conducting actual expansion readings on the affected member. On the other hand, condition assessment was conducted using cores extracted along the transversal direction, indicating the model's consistency, since expansion determined from the diagnosis falls within the range of average transversal expansion. It is worth mentioning that the data point representing the condition assessment in Figure 4.23 corresponds to the most-damaged column (i.e., 0.08% expansion) in order to demonstrate the "worst-case" scenario.

Figure 4.24 represents the multi-level chart developed by Sanchez et al. [42,43], which is based on the micro-mechanical coupling of ASR distress. Despite the fact that the above work was developed using specimens under "free expansion" conditions, recent work has shown that multi-level assessment may be used to appraise the condition of affected concrete as a function of "overall" microscopic damage (i.e., DRI number) regardless of confinement conditions [34]. According to the multi-level assessment of 45 MPa QC (i.e., Table 4.3 and purple chart – Figure 4.24), average ASR damage in 10 years is expected to result in losses of $\approx$ 15% in modulus of elasticity and $\approx$ 30% in tensile strength. Compressive strength loss at this stage is still expected to be negligible, indicating that the predicted damage is primarily a durability-related issue. However, rehabilitation measures are recommended at this stage to prevent future expansion, since smaller cracks are easier to mitigate and can prevent the development of significant future deformations.

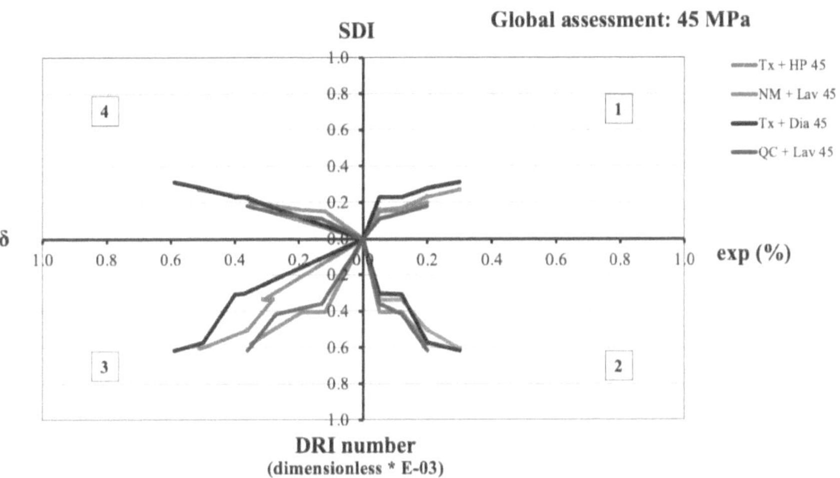

Figure 4.24: Spider plots as per [42,43] for 45MPa concrete containing different lithotypes.

4.7   Conclusion

This work presents a proposed model for the prediction of ASR-induced expansion curves of affected concrete structures in the field. The model is built upon the concept of correlating laboratory and field expansion based on the existing works of De Grazia et al. and Nguyen et al. Several aspects including the effects of geometry, casting direction, reinforcement and uniaxial loads were developed and incorporated into the simulation. Validations were successfully carried out using expansion records from field concrete to confirm the model's performance and accuracy. Finally, a case study was conducted to simulate the expansion of columns from the S.I.T.E. building at the University of Ottawa (Canada), which recently underwent a condition assessment. Findings from this study may be summarized as follows:

- Validation conducted using members from ASR exposure sites (Figure 4.14 and Figure 4.17) showed that the model matches the field results quite well. Thus, the proposed coefficients for anisotropy based on the uncoupled and coupled effects of confinements (i.e., stress and reinforcements) are quite effective in describing the behaviour of ASR-affected members on the structural level.

- The simulation of expansion determined for the case study for S.I.T.E. building columns indicated significant variation in expansion along different dimensions of the column due to the effect of anisotropy. Moreover, based on heat diffusion properties in concrete described in previous works, exposure conditions were optimized to conduct the prognosis of S.I.T.E. building columns, predicting a maximum average expansion of 0.10% in the columns.

- The coupled prognosis of the presented model and the multi-level assessment was shown to provide a well-rounded analysis of predicted damage, which could be quite useful in certain members where serviceability or even structural performance is of great significance.

4.8   References

[1]   B. Fournier, M.A. Bérubé, Alkali-aggregate reaction in concrete: A review of basic concepts and engineering implications, Can. J. Civ. Eng. 27 (2000) 167–191. https://doi.org/10.1139/l99-072.

[2]   A. Mohammadi, E. Ghiasvand, M. Nili, Relation between mechanical properties of concrete and alkali-silica reaction (ASR); a review, Constr. Build. Mater. 258 (2020) 119567. https://doi.org/10.1016/j.conbuildmat.2020.119567.

[3]   R.B. Figueira, R. Sousa, L. Coelho, M. Azenha, J.M. de Almeida, P.A.S. Jorge, C.J.R. Silva, Alkali-silica reaction in concrete: Mechanisms, mitigation and test methods, Constr. Build. Mater. 222 (2019) 903–931. https://doi.org/10.1016/j.conbuildmat.2019.07.230.

[4]   The Institution of Structural Engineers, Structural effects of alkali-silica reaction: Technical guidance on the appraisal of existing structures, 1992.

[5]   B. Fournier, M.A. Berube, K.J. Folliard, M. Thomas, Report on the Diagnosis, Prognosis, and Mitigation of Alkali-Silica Rection (ASR) in Transportation Structures, Austin, 2010.

[6]   M.A. Berube, B. Durand, D. Vézina, B. Fournier, Alkali-aggregate reactivity in Québec (Canada), Can. J. Civ. Eng. 27 (2000) 226–245. https://doi.org/10.1139/cjce-27-2-226.

[7]   M.A. Berube, J. Frenette, A. Pedneault, M. Rivest, Laboratory assessment of the potential rate of ASR expansion of field concrete, Cem. Concr. Aggregates. 24 (2002) 13–19. https://doi.org/10.1520/cca10486j.

[8]   N. Smaoui, B. Fournier, M.-A. Bérubé, B. Bissonnette, B. Durand, Evaluation of the expansion attained to date by concrete affected by alkali silica reaction. Part II: Application to nonreinforced concrete specimens exposed outside, Can. J. Civ. Eng. 31 (2004) 997–1011. https://doi.org/10.1139/l04-074.

[9]   M.A. Bérubé, N. Smaoui, B. Fournier, B. Bissonnette, B. Durand, Evaluation of the expansion attained to date by concrete affected by ASR - Part III: Application to existing structures, Can. J. Civ. Eng. 32 (2005) 463–479.

[10]   B. Godart, J.G.M. Wood, Appraising structures affected by the alkali-aggregate reaction, Proc. Inst. Civ. Eng. 169 (2016) 162–171. https://doi.org/10.1680/jcoma.15.00068.

[11] S. Multon, F.-X. Barin, B. Godart, F. Toutlemonde, Estimation of the Residual Expansion of Concrete Affected by Alkali Silica Reaction, J. Mater. Civ. Eng. 20 (2008) 54–62. https://doi.org/10.1061/(asce)0899-1561(2008)20:1(54).

[12] M.A. Hariri-Ardebili, V.E. Saouma, N.W. Hayes, A hybrid FE-based predictive framework for ASR-affected structures coupled with accelerated experiments, Eng. Struct. 234 (2021) 111709. https://doi.org/10.1016/j.engstruct.2020.111709.

[13] Y. Kawabata, K. Yamada, S. Ogawa, R. Martin, Y. Sagawa, S.E. Division, M. Division, S. Engineering, Correlation Between Laboratory Expansion and Field Expansion of Concrete : Prediction Based on Modified Concrete Expansion Test, 15th Int. Conf. Alkali-Aggregate React. (2016).

[14] R. Esposito, M.A.N. Hendriks, Literature review of modelling approaches for ASR in concrete: a new perspective, Eur. J. Environ. Civ. Eng. 23 (2019) 1311–1331. https://doi.org/10.1080/19648189.2017.1347068.

[15] V. Saouma, L. Perotti, Constitutive Model for Alkali-Aggregate Reactions, ACI Mater. J. (2007).

[16] V.E. Saouma, R.A. Martin, M.A. Hariri-Ardebili, T. Katayama, A mathematical model for the kinetics of the alkali-silica chemical reaction, Cem. Concr. Res. 68 (2015) 184–195. https://doi.org/10.1016/j.cemconres.2014.10.021.

[17] S. Poyet, A. Sellier, B. Capra, G. Thèvenin-Foray, J.-M. Torrenti, H. Tournier-Cognon, E. Bourdarot, Influence of Water on Alkali-Silica Reaction: Experimental Study and Numerical Simulations, J. Mater. Civ. Eng. 18 (2006) 588–596. https://doi.org/10.1061/(asce)0899-1561(2006)18:4(588).

[18] B. Capra, J.P. Bournazel, Modeling of induced mechanical effects of alkali-aggregate reactions, Cem. Concr. Res. 28 (1998) 251–260. https://doi.org/10.1016/S0008-8846(97)00261-5.

[19] C. Larive, Apports combinés de l'expérimentation et de la modélisation à la compréhension de l'Alcali-réaction et de ses effets mécaniques, in: 1998.

[20] F.-J. Ulm, O. Coussy, L. Kefei, C. Larive, Thermo-Chemo-Mechanics of ASR Expansion in Concrete Structures, J. Eng. Mech. 126 (2000) 233–242.

[21] M.T. De Grazia, N. Goshayeshi, R. Gorga, L.F.M. Sanchez, A.C. Santos, D.J. Souza, Comprehensive semi-empirical approach to describe alkali aggregate reaction (AAR) induced expansion in the laboratory, J. Build. Eng. 40 (2021). https://doi.org/10.1016/j.jobe.2021.102298.

[22] T.N. Nguyen, L.F.M. Sanchez, J. Li, B. Fournier, V. Sirivivatnanon, Correlating ASR-induced expansion from short term laboratory testings to long-term field performance: a novel semi-empirical model, University of Technology Sydney, 2019.

[23] N. Smaoui, M.-A. Bérubé, B. Fournier, B. Bissonnette, Influence of Specimen Geometry, Orientation of Casting Plane, and Mode of Concrete Consolidation on Expansion Due to ASR, 2004.

[24] S.H. Diab, A.M. Soliman, M.R. Nokken, Effect of triggering material, size, and casting direction on ASR expansion of cementitious materials, Constr. Build. Mater. 269 (2021). https://doi.org/10.1016/j.conbuildmat.2020.121323.

[25] S. Multon, F. Toutlemonde, Effect of applied stresses on alkali-silica reaction-induced expansions, Cem. Concr. Res. 36 (2006) 912–920. https://doi.org/10.1016/j.cemconres.2005.11.012.

[26] E.R. Giannini, Evaluation of Concrete Structures Affected by Alkali-Silica Reaction and Delayed Ettringite Formation, (2012). https://doi.org/10.13140/2.1.2897.9525.

[27] N. Sinno, M.H. Shehata, Effect of sample geometry and aggregate type on expansion due to alkali-silica reaction, Constr. Build. Mater. 209 (2019) 738–747. https://doi.org/10.1016/j.conbuildmat.2019.03.103.

[28] N. Smaoui, B. Bissonnette, M.A. Bérubé, B. Fournier, Stresses induced by alkali-silica reactivity in prototypes of reinforced concrete columns incorporating various types of reactive aggregates, Can. J. Civ. Eng. 34 (2007) 1554–1566. https://doi.org/10.1139/L07-063.

[29] T. Garcia, M.S. Mirza, Effect of reinforcement on aar expansion in concrete, Proc. Int. Conf. Appl. Codes, Des. Regul. (2005) 271–279.

[30] J. Liaudat, I. Carol, C.M. López, V.E. Saouma, ASR expansions in concrete under triaxial confinement, Cem. Concr. Compos. 86 (2018) 160–170.

https://doi.org/10.1016/j.cemconcomp.2017.10.010.

[31] B.P. Gautam, D.K. Panesar, S.A. Sheikh, F.J. Vecchio, Multiaxial expansion-stress relationship for alkali silica reaction-affected concrete, ACI Mater. J. 114 (2017) 171–184. https://doi.org/10.14359/51689490.

[32] H. Aryan, B. Gencturk, M. Hanifehzadeh, J. Wei, ASR Degradation and Expansion of Plain and Reinforced Concrete, Struct. Congr. 2020 - Sel. Pap. from Struct. Congr. 2020. (2020) 303–315. https://doi.org/10.1061/9780784482896.029.

[33] A. Allard, S. Bilodeau, F. Pissot, B. Fournier, J. Bastien, B. Bissonnette, Expansive behavior of thick concrete slabs affected by alkali-silica reaction (ASR), Constr. Build. Mater. 171 (2018) 421–436. https://doi.org/10.1016/j.conbuildmat.2018.03.159.

[34] A. Zahedi, C. Trottier, L.F.M. Sanchez, M. Noël, Microscopic assessment of ASR-affected concrete under confinement conditions, Cem. Concr. Res. 145 (2021) 106456. https://doi.org/10.1016/j.cemconres.2021.106456.

[35] M.H. Shehata, M.D.A. Thomas, Effect of fly ash composition on the expansion of concrete due to alkali-silica reaction, Cem. Concr. Res. 30 (2000) 1063–1072. https://doi.org/10.1016/S0008-8846(00)00283-0.

[36] U. Costa, T. Mangialardi, A.E. Paolini, Minimizing alkali leaching in the concrete prism expansion test at 38 °C, Constr. Build. Mater. 146 (2017) 547–554. https://doi.org/10.1016/j.conbuildmat.2017.04.116.

[37] J. Lindgård, M.D.A. Thomas, E.J. Sellevold, B. Pedersen, Ö. Andiç-Çakir, H. Justnes, T.F. Rønning, Alkali-silica reaction (ASR) - Performance testing: Influence of specimen pre-treatment, exposure conditions and prism size on alkali leaching and prism expansion, Cem. Concr. Res. 53 (2013) 68–90. https://doi.org/10.1016/j.cemconres.2013.05.017.

[38] N. Sinno, M. Shehata, Understanding the factors affecting expansion due to alkali-silica reaction in concrete, in: 7th Int. Mater. Spec. Conf. 2018, Held as Part Can. Soc. Civ. Eng. Annu. Conf. 2018, 2019: pp. 394–404.

[39] Kingston Outdoor Exposure Site for ASR - 27 Year Update, Kingston, 2018.

[40] B. Fournier, J.H. Ideker, K.J. Folliard, M.D.A. Thomas, P.C. Nkinamubanzi, R. Chevrier,

Effect of environmental conditions on expansion in concrete due to alkali-silica reaction (ASR), Mater. Charact. 60 (2009) 669–679. https://doi.org/10.1016/j.matchar.2008.12.018.

[41] R.D. Hooton, C. Rogers, C.A. MacDonald, T. Ramlochan, Twenty-year field evaluation of alkali-silica reaction mitigation, ACI Mater. J. 110 (2013) 539–548. https://doi.org/10.14359/51685905.

[42] L.F.M. Sanchez, B. Fournier, M. Jolin, D. Mitchell, J. Bastien, Overall assessment of Alkali-Aggregate Reaction (AAR) in concretes presenting different strengths and incorporating a wide range of reactive aggregate types and natures, Cem. Concr. Res. 93 (2017) 17–31. https://doi.org/10.1016/j.cemconres.2016.12.001.

[43] L.F.M. Sanchez, T. Drimalas, B. Fournier, D. Mitchell, J. Bastien, Comprehensive damage assessment in concrete affected by different internal swelling reaction (ISR) mechanisms, Cem. Concr. Res. 107 (2018) 284–303. https://doi.org/10.1016/j.cemconres.2018.02.017.

[44] L.F.M. Sanchez, Contribution to the assessment of damage in aging concrete infrastructures affected by alkali-aggregate reaction, PhD., (2014) 377.

[45] M. Bérubé, B. Durand, D. Vézina, B. Fournier, Alkali – aggregate reactivity in Québec (Canada), 245 (2000) 226–245.

[46] C. Rogers, R.D. Hooton, J. Ryell, M.D.A. Thomas, Alkali-aggregate reactions in Ontario Alkali – aggregate reactions in Ontario, NRC Publ. Arch. (2000).

[47] V.E. Saouma, Diagnosis & prognosis of AAR affected structures : state-of-the-art report of the RILEM Technical Committee 259-ISR, Springer, Cham, Switzerland, 2021.

[48] CSA S6-14, Canadian highway bridge design code, Missisauga, 2014.

[49] F. Rajabipour, E. Giannini, C. Dunant, J.H. Ideker, M.D.A. Thomas, Alkali-silica reaction: Current understanding of the reaction mechanisms and the knowledge gaps, Cem. Concr. Res. 76 (2015) 130–146. https://doi.org/10.1016/j.cemconres.2015.05.024.

[50] R.A. Deschenes, E.R. Giannini, T. Drimalas, B. Fournier, W.M. Hale, Effects of moisture, temperature, and freezing and thawing on Alkali-Silica reaction, ACI Mater. J. 115 (2018) 575–584. https://doi.org/10.14359/5170219.

[51] R. V. Gorga, L.F.M. Sanchez, B. Martín-Pérez, FE approach to perform the condition

assessment of a concrete overpass damaged by ASR after 50 years in service, Eng. Struct. 177 (2018) 133–146. https://doi.org/10.1016/j.engstruct.2018.09.043.

This research was conducted with the objective of conducting the diagnosis and prognosis of the S.I.T.E. building at the University of Ottawa which shows signs of ASR development after nearly 20 years in service. The multi-level assessment protocol was central in the evaluation of current and future predicted ASR damage in affected columns on the structural level (i.e., durability and serviceability) through the coupling of microscopic and mechanical distress. Diagnosis was carried out on rotunda columns through a two-step procedure: 1) visual inspection, and 2) core-based evaluation through the use of microscopic and mechanical testing methods (i.e., multi-level assessment). Prognosis was conducted using a novel ASR semi-empirical model developed to simulate the induced expansion in the ASR-affected columns. Validation conducted using expansion data from ASR exposure sites revealed that existing semi-empirical models for the prediction of the chemo-mechanical behaviour of ASR-affected concrete are applicable for affected members on the structural level. Moreover, the mathematical relations developed for the model successfully captured anisotropic behaviour of induced expansion. In combination with the multi-level assessment, the model was able to roughly predict the mechanical property reduction of the column in the next 10 years. Findings from the above work could be summarized as follows:

- Visual inspection revealed surface damage (i.e., map cracking, spalling, discoloration, and gel exudations) indicating potential ASR. The presence of ASR, which was validated through petrographic analysis conducted in natural and polarized light and supplemented by SEM, characterized the reactive aggregate as a dolomitic limestone. Afterwards, multi-level assessment was able to identify damage degrees ranging from marginal (i.e., 0.03% and 0.05%) to moderate (0.08%) ASR damage. Moreover, observations from DRI analysis showed an increasing amount of cement paste cracks originating in the bulk paste as expansion increased, suggesting that the damage mechanism of FT is developing increasingly in the more damaged columns. High FT damage could result in future problems with the structural safety of columns, since FT is characterized by damage in the bulk cement paste, which could effectively reduce concrete strength and thus, the structural capacity of the columns. At the moment, the durability and serviceability of the columns is not of immediate concern, yet this could change quickly in the short-term (i.e., next 5 to 10 years), especially in the likely event that FT damage will continue to increase. Overall, this paper shows that the multi-level assessment is a quite reliable protocol for the appraisal and management of aging infrastructures. Moreover, a relation was found between average crack widths (recorded during visual inspection) and the damage degree as per the multi-level assessment. Hoever, comparisons of multi-level assessment with visual inspection showed that the latter is unable to characterize damage degree in its current form as it is quite sensitive to exposure conditions (i.e., sun/wind exposure, wetting and drying, etc.).

- The average expansion of the S.I.T.E. columns was predicted to be $\approx 0.16\%$ in 10 years, based on the model simulation, taking into account the most severely diagnosed column (0.08%). By combing the model results with multi-level assessment, reductions of $\approx 30\%$ and $\approx 48\%$ in stiffness and tensile strength, respectively, are predicted in the most severely damaged columns. However, the model showed that significant anisotropy due to column geometry and confinement could result in drastic differences between directional expansion, and thus highly non-homogeneous mechanical behavior.

- Validations of the above model showed that considerations for the chemo-mechanical and structural effects of ASR expansion were quite accurate, although few limitations could be identified. In its current form, the model is unable to account for reductions in volumetric expansion due to high stress states. Therefore, in rare circumstances, simulation could be misleading. Moreover, anisotropy due to stress confinement incorporated in the model is limited to uniaxial loads, and unable to address more complex stress states that are normally found in structural members.

Based on the above findings and limitations, some recommendations for future research were identified as follows:

- Previous research, including this study, has shown that map-cracking techniques used for visual inspection (e.g., cracking index) are not consistent with testing protocols for core-based evaluations. However, the correlation between crack width and the multi-level assessment shows that map-cracking techniques may have the potential for diagnosing concrete damage yet must undergo comprehensive research to optimize the analytical methodology.

- The geometrical component of the ASR expansion model was developed using test data reported from previous studies, where expansion was measured along multiple directions per specimen. However, due to the scarcity of data, it was not possible to examine whether there are any differences in anisotropy based upon the type/nature of reactive aggregates. Moreover, the current form of the model is only able to simulate expansion for rectangular members, and unable to account for anisotropy due to member size. Thus, a through experimental study incorporating a range of geometrical proportions, sizes, and several aggregate lithotypes will make the model more comprehensive.

The loading conditions for the SITE building based on the engineering drawings are detailed in the table below. Given the two-storey structure of the building, both floor and roof loads were considered. As per Canadian design standards (CSA A23.3, Cl. 9.3), the critical load combination was found to be $1.25D + 1.5L$, where D and L represent dead and live loads respectively. Accordingly, Appendix B presents modeling results using SAP2000 based on approximate frame analysis.

	Floor loads		Roof loads	
	Type	Value	Type	Value
Dead loads	Toppings	1.35 kPa	Membrane	0.75 kPa
	H.C. slabs	3.86 kPa	Sloped topping	2.00 kPa
	Partitions	1.00 kPa	Topping	1.35 kPa
	Mechanical	0.25 kPa	H.C. slabs	3.86 kPa
	Finishes	0.25 kPa	Mechanical & finishes	0.25 kPa
	Total	6.71 kPa	Total	8.21 kPa
Live loads	Total	4.80 kPa	Total	2.30 kPa

Appendix B: SITE Building Structural Modeling

Axial Loads Acting on Columns (kN)

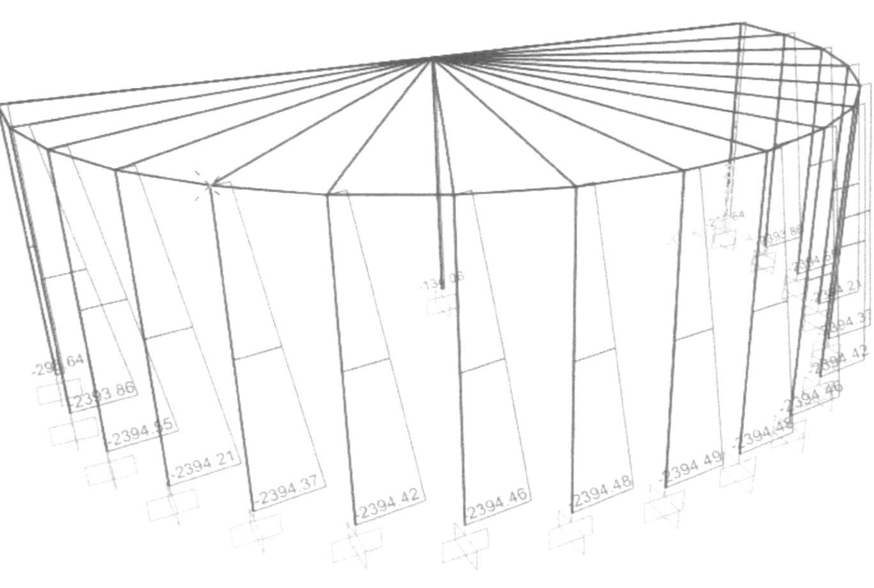

Bending Moments Acting on Beams (kNm)

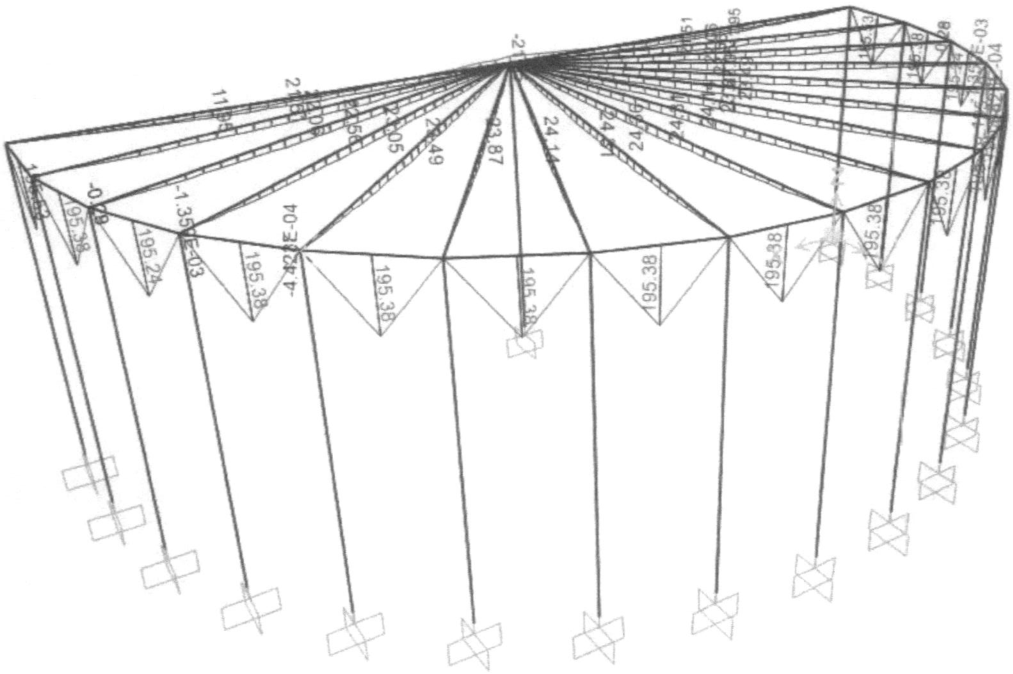

Bending Moments Acting on Columns (kNm)

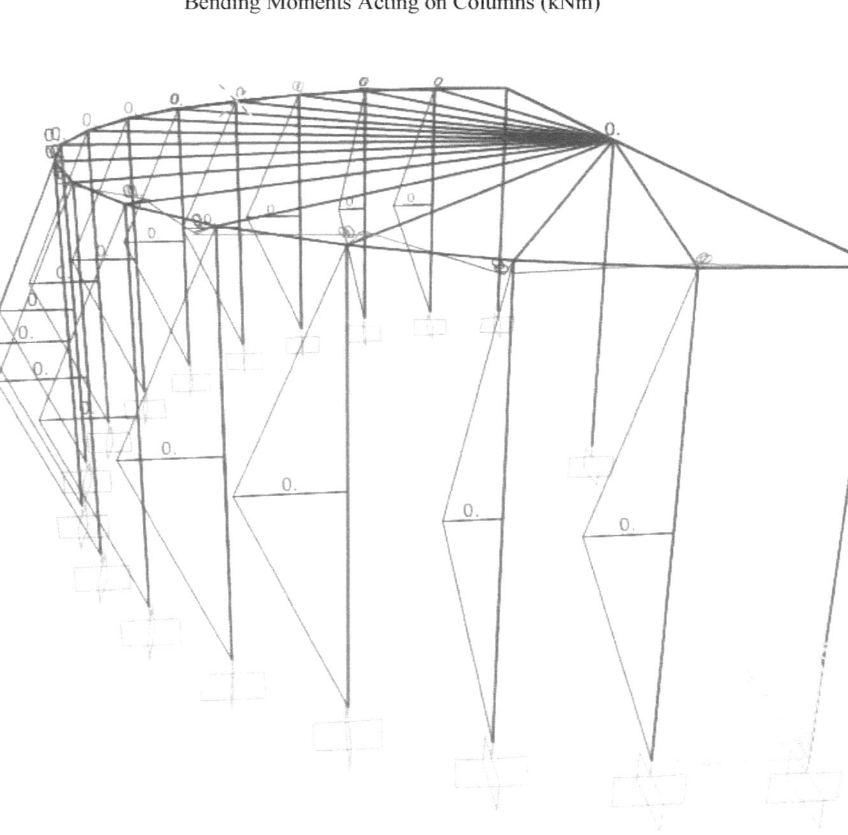